BASIC NUCLEAR PHYSICS FOR MEDICAL PERSONNEL

BASIC NUCLEAR PHYSICS FOR MEDICAL PERSONNEL

By

HERBERT L. JACKSON, Ph.D.

Radiation Physicist
Professor of Radiology
University of Iowa
Iowa City, Iowa

CHARLES C THOMAS • PUBLISHER
Springfield • Illinois • U.S.A.

Published and Distributed Throughout the World by
CHARLES C THOMAS • PUBLISHER
BANNERSTONE HOUSE
301–327 East Lawrence Avenue, Springfield, Illinois, U.S.A.

This book is protected by copyright. No part of it may be reproduced in any manner without written permission from the publisher.

© 1973, by CHARLES C THOMAS • PUBLISHER

ISBN 0-398-02663-7

Library of Congress Catalog Card Number: 72-88471

With THOMAS BOOKS *careful attention is given to all details of manufacturing and design. It is the Publisher's desire to present books that are satisfactory as to their physical qualities and artistic possibilities and appropriate for their particular use.* THOMAS BOOKS *will be true to those laws of quality that assure a good name and good will.*

Printed in the United States of America

BB-14

To my parents
ROY ALVIN JACKSON
and his wife
GRACE STOOPS

PREFACE

ALTHOUGH SEVERAL GOOD BOOKS on nuclear physics have been published in recent years for the physician specializing in nuclear medicine, suitable instructional literature for the medical trainee having a meagre scientific background is still sparse. It is our hope that the present volume will fill that gap.

When planning a book of this sort, a number of questions arise, none of which has a single, clear, indisputable answer. For example: (1) For whom, exactly is one writing? (2) Should one strive for completeness or deliberately limit the text to selected topics? (3) To what extent should one stress the quantitative aspect of the subject (or more bluntly) mathematics? (4) Should the treatment be centered upon basic theory or upon scientific applications? The choice of answers to such questions determines the nature of the work.

Broadly stated, the possible readers of this book include all medical personnel who expect to be concerned directly or indirectly with nuclear medicine in the future. If the use of radioactive materials continues to increase as rapidly as it has in the recent past, that group will eventually include almost everyone working in a hospital or clinic. Perhaps even the janitors will need to know a few basic rules of health physics, be able to recognize radioactive materials, and make simple radiation surveys.

Primarily however this book is designed as a classroom text for student technicians in nuclear medicine and radiology. We hope, though, that the physician entering a residency in nuclear medicine will find it a simple, easily readable refresher of elementary theory. Perhaps specialists in other medical disciplines such as internal medicine and pathology will also find it helpful in acquainting themselves with the field.

For a subject as extensive as nuclear physics completeness is of course impossible. Besides, as we are convinced that it is far better to master a little knowledge than to cram the mind with countless details, the subject matter of this book has been limited to those basic concepts relevant to present and probable future applications of nuclear medicine. Consequently one will find no mention of mesons, strangeness, nor non-conservation of parity in this volume, important as those topics may be.

We have decided to treat basic principles rather than specific applications for two reasons. First, information about any particular piece of equipment, pharmaceutical, or procedure quickly becomes dated. More important, we firmly believe that he who knows the fundamentals of his trade is better able to make an acceptable judgment when the need arises than his robotized brother. He can quickly adapt to new methods and equipment, whereas the latter, who knows only how to operate a particular machine or perform a certain technique must be retrained whenever any change is made.

In the first part of the book considerable space has been given to the structure of bulk matter and of the atom. The purpose of this material is to lead the reader by easy steps from the world of everyday experience to the mysterious realm of the atomic nucleus. It is important that the student have a clear notion of the proper place and function of the atomic nucleus in Nature.

Since elementary nuclear physics is one of the more descriptive branches of physics, it has been no problem to keep the mathematical content of the book to a minimum. The mind-wrenching computations demanded of researchers and theoreticians do not concern the beginner.

Finally I would like to express my appreciation to Dr. Latourette for his interest and encouragement and to the many students who—in all likelihood—have taught the teacher more than he taught them.

INTRODUCTION

OF ALL THE ancient peoples, the Greeks were most fond of teasing themselves with unanswerable questions. One of the more famous was: Suppose one were to take a sample of any substance, say wood, stone, or metal, and cut it into smaller and smaller pieces. Could this process go on forever—given the necessary patience and equipment—or would one finally reach particles so hard that no instrument whatever could break them down further? While each view had its supporters, most thinkers believed that there must be a stopping point; everlasting, unchangeable particles must be the ultimate stuff of matter. The Greeks called these supposed granules the "indivisible ones" (τὰ ἄτομα) whence the modern word *atom*. To anticipate, it may be said that while we use the word *atom*, both of the historical viewpoints are equally wrong; matter is not infinitely divisible, nor does it consist of ultimate, unchangeable particles.

The Greeks made no progress in answering the question, partly because they lacked the tools, but mainly because they were not really interested in trying. They found it much more fun to argue an issue than to investigate it. Had they been research-minded, they could have contrived the necessary instrumentation from the materials at hand. In so doing they would have founded the sciences of physics and chemistry two millenia earlier than actually occurred.

To illustrate the atomic concept, imagine that you are breaking up a piece of sandstone. The task proceeds easily until you reach the individual grains. Here you must stop. The analogy is not perfect since by using sufficient force a sand grain can be shattered. The philosophic atom on the other hand was supposed to be eternal, completely immune to any form of wear or breakage.

The first experimental evidence that matter consists of submicroscopic particles was obtained by the chemists and alchemists of the late eighteenth and early nineteenth centuries. They immediately assumed these particles to be the atoms foreseen by the Greeks and hence so called them. Soon however it was obvious that many of these particles were aggregates of still smaller ones. So the originally discovered particles were renamed *molecules*, and the smaller ones, *atoms*.

It was taken for granted that all bulk matter consists of molecules which are in turn ensembles of atoms. The atoms themselves were supposed to have either existed throughout all eternity or to have been divinely created in the Beginning. These assumptions are both wrong as we shall see; Nevertheless, they were so intellectually satisfying and so successful in accounting for most of the then-known experimental facts that they became unquestioned dogma.

There was some evidence to the contrary that received less attention than it deserved. For one thing, about eighty different kinds of atoms had been discovered, a most illogical number. One would have expected either very few—say, about half a dozen—or perhaps infinitely many. Furthermore, the different kinds of atoms cluster into families with similar properties. Such behavior suggests that the individual members of such families have similar structure; yet the very notion of structure contradicts the atomic hypothesis, for anything that has structure is divisible, at least conceptually.

In spite of these hints, many were severely shocked to discover that the atom can be taken apart. Indeed, once one knows how, he can knock chips off of an atom with ridiculous ease. A tiny spark from a battery rips electrons off the atoms in its path; dissolving many a common substance in water may remove electrons from its atoms. With more drastic methods the innermost core of any atom can be broken.

Not only is it possible to disassemble any atom; many disintegrate of their own accord. Some emit electrons and others helium nuclei. Still others split apart into two nearly equal halves.

These findings triggered off a frenzy of research. During the first third of this century, it was proved that every atom is an assembly of protons, electrons, and (except for ordinary hydrogen) neutrons. The properties of these subatomic particles will be discussed later. The number of protons in an atom determines its chemical properties. The number of neutrons is usually equal to or somewhat greater than the number of protons. The exact number may vary for different atoms of the same element. The number of electrons in an atom also varies but is usually about equal to the number of protons. It was further discovered that the protons and neutrons are concentrated in a tiny sphere in the center of the atom while the electrons are dispersed throughout the entire atomic volume.

Are the electrons, protons, and neutrons the atoms of the philosophers? Are these the changeless, unbreakable entities of the universe? No, they are not; for while no one has succeeded in cutting any of these particles into smaller ones, they not only have internal structure but may also be transformed into each other or into non-material energy. Conversely, energy can be materialized into electrons, protons, and neutrons as well as a variety of less common particles. No permanent, immutable particles are known nor is there reason to suppose that such exist. To misquote Shakespeare, philosophers have dreamed of many things of which no trace can be found in heaven or earth.

The story of the atom shows how futile it is to speculate. The assumption that matter had to be either continuous or atomic was and is perfectly reasonable, but like all *either-or* logic overlooks the possibility of third, fourth, or fifth choices.

Physics offers an even better example of the insidious power of *either-or* logic. Newton suggested that a beam of light is a stream of fast-moving particles; at about the same time Huyghens argued that it is a train of waves. From that moment on everyone was obsessed by the notion that light must consist *either* of waves *or* of particles. A century of misguided research was devoted to resolving the problem.

Now we know that the basic units of light, called *photons*, are uniquely different from anything else in the universe. They resemble particles from certain points of view, waves from others, and neither from still others. Perhaps we can compare the photon to the coyote. Sometimes he looks and acts like a dog; at other times a wolf. He is not a dog or a wolf, however, but a distinct species in his own right.

CONTENTS

	Page
Preface	vii
Introduction	ix

Chapter

I.	Force	3
II.	Work and Energy	12
III.	Units and Constants of Microscopic Physics	23
IV.	Fundamental Particles	28
V.	Elements, Nucides, Etc.	36
VI.	The Structure of Bulk Matter	41
VII.	Atomic Structure	47
VIII.	Nuclear Structure	65
IX.	Radioactive Decay	75
X.	Nuclear Decay	83
XI.	Natural Radioactivity	98
XII.	Artificial Radioactivity	102
XIII.	The Interaction of Radiation with Matter	110
XIV.	Dose and Exposure	131

Index	143

BASIC NUCLEAR PHYSICS FOR MEDICAL PERSONNEL

Chapter 1 FORCE

DEFINITION AND UNITS

IF LEFT ALONE, any material object at rest will remain at rest and if in motion, will continue that motion in a straight line at constant speed. Any other behavior can be brought about only by applying what is called a *force,* or in everyday language, a *push* or a *pull.* By a proper choice of forces, a stationary object can be made to move, and moving objects can be stopped or compelled to change their speed or direction. If there were no forces, the universe as we know it would not exist as the individual particles composing it would have long ago dispersed into outermost space—unless of course they were absolutely stationary at the outset.

To the casual eye, forces appear to be of two kinds, *contact forces* and *forces-at-a-distance.* When one car pushes another bumper-to-bumper, or one pulls the other with a chain, contact forces come into play. Examples of forces at a distance include weight, which is the gravitational force that the earth exerts on any object near it, the force of a magnet on a piece of iron, the force that a briskly rubbed comb exerts on pieces of lint and so on. In the field of physics this distinction between contact and remotely acting forces turns out to be unimportant. In fact all forces are of the latter type. Contact forces appear to be so because they come into play only at distances smaller than the eye can perceive. The so-called contact forces are really due to the electrostatic attraction and repulsion between atoms and molecules that are quite near one another.

The study of forces is somewhat more complicated than it need be because of the many different units in common use. Even worse, units of mass are commonly used to designate forces, a practice that is admittedly convenient but dangerously confusing. The beginner should concern himself only with the *dyne,* the *newton,* and the *pound.* When these are mastered (as well as the concept of force itself) he can then learn some of the other units if he has need of them.

The dyne, like most physical units, is defined in terms of a specific experimental situation. Let a material object (such as a steel ball) having a mass of one gram be placed at rest in a force-free region of space. Then, as we have said, it would remain there permanently unless disturbed. Now apply a force of constant strength and direction to it for one second. If thereafter the object is moving with a speed of exactly one centimeter per second, the amount of force applied was one dyne. While this definition presupposes an experimental set-up that cannot be achieved in practice, there are indirect ways of accomplishing the same thing.

The *newton* is defined in the same way as the dyne except that the mass of the object to be moved is one kilogram and the final speed is one meter per second. The result of these changes is that the newton is 100,000 times larger than the dyne.

The dyne, which is the older of the two units, is too awkward in size for general use. For the tasks of everyday living, millions of dynes are required. Just to lift an average-size book requires about half a million dynes. The newton is much more convenient. It is the force required to lift a 98 gram mass on the earth's surface. It is also very nearly 2/9 of a pound.

To us living in an English speaking country the pound is quite familiar. Unfortunately it is indifferently used both as a measure of mass and of force. To illustrate the difference, consider a pound of butter. If we think of the amount of butter present, we are dealing with mass; if we think of the amount of effort required to lift it, we are dealing with force.

KINDS OF FORCE

According to our present knowledge there are four distinct kinds of force, all alike in that they exert a push or a pull upon the body to which they are applied, but completely different in origin. They also differ in their strengths and in the way they vary with distance. While it is suspected that there is a common mechanism responsible for all of them, none has so far been found. These four forces are:

The gravitational force
The weak force
The electromagnetic force
The nuclear force

They are listed in the order of increasing strength. The term *electromagnetic force* includes both the electrostatic and the magnetic force since they have a common origin in spite of their seeming differences.

THE GRAVITATIONAL FORCE

The gravitational force is the most familiar as it is the one that draws us and all surrounding objects to the earth's surface.

Let us digress for a moment to define a word often misused and misunderstood—*weight*. Properly speaking, weight is the gravitational force which the earth exerts on another body at or near its surface. In this space age it is also correct to speak of the force which the moon, a planet, or other heavenly body exerts on an object in its vicinity as weight. Thus one can say that a man on the moon weighs only one sixth as much as on earth and so on.

Note particularly the distinction between mass and weight. Mass is the quantity of matter that an object contains and does not change unless substance is actually added or removed.*
Thus a man whose mass is 70 kg on earth would also have a mass of 70 kg on the moon, other things remaining equal. His weight however would be less because the moon, being a

---

* The relativistic variation of mass with energy content is an exception that is not relevant to the present discussion.

smaller object than the earth cannot exert as large a force on the objects near it.

Although the gravitational force appears quite strong, it is, in fact, by far the weakest of the four. Its seeming strength is due to the enormous mass of the earth. The gravitational force between a couple of *man-sized* objects is almost impossible to detect. For bodies of atomic size, the gravitational force is billions upon billions of times too weak to have any observable effect.

The magnitude of the gravitational force between two objects is expressed by the formula:

$$F = G \frac{M_1 \times M_2}{d^2}$$

Where F is the force, M_1 and M_2 are the masses of the two interacting objects, d is the distance between them, and G is the gravitational constant. If the masses are expressed in kilograms, the distance in meters, the force in newtons, then the value of G is:

$$-6.673 \times 10^{-11} \text{ nt--m}^2/\text{kg}^2$$

The minus sign is present to satisfy the convention that attractive forces are regarded as negative and repulsive forces as positive.

PROBLEMS

The following data are needed for the problems:

Mass of the earth	5.983×10^{24} kg
Mass of the sun	1.971×10^{30} kg
Mass of the moon	7.347×10^{22} kg
Sun-earth distance	1.495×10^{8} km
Moon-earth distance	3.844×10^{5} km
Diameter of the moon	3.476×10^{3} km
Diameter of the earth	1.276×10^{4} km

1. What is the gravitational force between the sun and the earth?

2. What is the force of attraction between the earth and the moon?

3. How much less will an object weigh on the moon than on the earth?

4. What is the gravitational force between two cannon balls placed a meter apart if each has a mass of ten kg?

5. At what point between the earth and the moon are the gravitational forces of the two bodies on a given object equal and opposite?

THE WEAK FORCE

Since the weak force is much stronger than the gravitational force, the name appears inappropriate. It was coined by those physicists who study nuclear and elementary particle interactions; for these people the gravitational force just does not exist—professionally speaking.

Since the nuclear force and the electromagnetic forces account for most atomic and nuclear processes, the weak force was unrecognized for a number of years. Some phenomena, however, particularly beta decay, require a third type of interaction, which experiment shows to be about a trillion times weaker than the electromagnetic forces. Little is known about the weak force apart from its effects.

THE ELECTROMAGNETIC FORCE

There appear to be two kinds of electromagnetic forces, the electrostatic force and the magnetic force. They were both known in antiquity but were regarded as amusing curiosities until Tudor times when Gilbert published his famous and scholarly work *De Magnete*. Even then these forces were not studied intensively until the late eighteenth and early nineteenth centuries. It was first thought that the electrostatic and the magnetic force were identical; when that viewpoint became untenable, they were thought to be unrelated; now we know that they are different aspects of a more general phenomenon: the interaction between electrical charges.

If electrical charges are separated from one another but are not moving, the force is purely electrostatic; if the charges are not spatially separated but are in motion, the force is purely magnetic. This latter condition exists in a metallic conductor where a continuous stream of electrons flow past the stationary positively charged atomic nuclei. Since in every region of the conductor the concentration of positive and negative charges are equal, no net electrostatic forces exist. If the charges are both separated and moving, a more complicated force occurs which has both electrostatic and magnetic components.

The nature of the electromagnetic force depends partly upon the observer. Imagine a laboratory with two electrically charged objects in it. To an observer in that laboratory who is not moving relative to either charge, there will appear to be a purely electrostatic force between them. Now suppose that someone outside, who is passing by in a rapidly moving vehicle has with him an instrument capable of detecting electric and magnetic forces. He will of course also detect an electrostatic force because the two charges are in fact separated. In addition however, he will detect a magnetic force because (from his point of view) the charges are in motion.

The expression for the electrostatic force has exactly the same mathematical structure as that for the gravitational force:

$$F = \left(\frac{1}{4\pi\epsilon}\right) \frac{Q_1 \times Q_2}{d^2}$$

where Q_1 and Q_2 are the two interacting charges, d is the distance between them, and $\left(\frac{1}{4\pi\epsilon}\right)$ is the constant of proportionality. If the force is to be expressed in newtons, the charges in coulombs, and the distance in meters, then the value of the constant of proportionality is:

$$\frac{1}{4\pi\epsilon} = 8.988 \times 10^9 \doteq 9 \times 10^9$$

The reason why the constant of proportionality is written in so complicated a fashion is that a number of frequently used formulas are thereby simplified. The value of the constant is such as to make the common electrical units, the *volt*, the *am-*

pere, and the *ohm* consistent with the metric units of measure, the *meter,* the *kilogram,* and the *second.*

Notice that if both charges are of the same sign, the force is positive and hence repulsive; if the charges are of opposite sign, the force is negative and hence attractive.

The expression for the magnetic force is much more complicated. The reason is that the current at any given point has not only a certain magnitude but a definite direction as well. Furthermore, a current cannot be restricted to a point as can a charge but must flow over a path of some length. The formula for the magnetic force must therefore take account of the strength of the current, the length and shape of the path, and the direction of the current at each point. In vector notation the formula looks like this:

$$\vec{F}_{mag} = \frac{\mu l_p l_s}{4\pi} \iint \frac{\vec{da_p} \times \vec{da_s} \times \vec{r_{sp}}}{r_{sp}^3}$$

We will not attempt to use it for any calculations.

PROBLEMS

1. The distance between the electron and proton in the hydrogen atom is approximately one-half an angstrom, that is, 5.0×10^{-11} m. The magnitude of the charge on each is 1.6×10^{-19} coulomb. What is the force on each?

2. Suppose that all the atoms of a gram of hydrogen were separated into electrons and protons. The number of atoms involved is 6.02×10^{23}. The electrons are placed in one container and set at the north pole. The protons are placed in another container and placed at the south pole. If the diameter of the earth is 1.276×10^4 km, what is the force between the two containers?

3. What is the gravitational force between the electrons and protons mentioned in problem 2? What is the ratio between the gravitational and the electrostatic forces involved? The mass of the proton is 1.672×10^{-27} kg and that of the electron is 9.107×10^{-31} kg.

THE NUCLEAR FORCE

The nuclear force is the strongest of the four types found in nature. It is a little over one hundred times stronger than the electromagnetic force and much more complex. No one has succeeded in isolating all the factors upon which it depends nor in writing a formula that describes it accurately. The following facts about it however go far to explain many of the features of nuclear behavior:

1. The nuclear force is effective only at distances of about 10^{-13} m; that is to say, distances considerably less than the diameter of most atomic nuclei.
2. The nuclear force acts only between pairs of neutrons and/or protons. Electrons and neutrinos are not affected.
3. The strength of the interaction is about the same whether the interacting particles are both neutrons, both protons, or one of each.

Because of its extremely short range, the nuclear force cannot simultaneously affect a large number of particles, a fact of paramount importance for understanding nuclear phenomena. Each proton and each neutron in any nuclear structure is held in place only by forces acting between it and the immediately neighboring particles. The situation is somewhat like that prevailing in a brick wall where each brick is held in place only by the mortar attaching it to the six adjacent bricks.

Although weaker than the nuclear force, the electrostatic force is operative at all distances, and moreover, since the only charges within the atomic nucleus are positive, it is purely disruptive. Consequently successively larger nuclei are progressively less stable, for while the attractive force tending to hold a particular proton in place is about the same for all nuclei (except the very smallest) the repulsive electrostatic force tending to drive it out increases with nuclear size. If the total number of protons in a nucleus is greater than 82—meaning that is beyond lead in the periodic table—that nucleus cannot be permanently stable but will eventually be reduced to a smaller structure either by alpha emission or fission.

While it is clear from the foregoing that the number of protons within an atomic nucleus must be limited, one can reasonably ask why there are not huge nuclei containing hundreds or thousands of neutrons. The answer is that the earlier mentioned weak force will transform one neutron after another into a proton until a counteractive force inhibits the action. In the transformation an electron and an antineutrino are also produced, but these escape and play no further role in the behavior of the nucleus. The only effective way of blocking neutron decay is to surround it with a strong positive charge, that is to say, a large number of protons.

Suppose we begin with a nucleus that contains a large excess of neutrons. If the total number of particles present is less than about 200, neutrons will turn into protons one by one until the total number of protons is equal to one third to one half of the total. Thereafter the nucleus will be a stable structure capable of lasting indefinitely. In larger neutron-rich nuclei the transformation of neutrons into protons continues until the electrostatic force is sufficient to blow the assembly apart.

Chapter II WORK AND ENERGY

WORK

IN OUR EVERYDAY LIFE it is usually impossible to bring about any change in our surroundings without expending some effort. This fact is not an accident of human existence but is basically true of nature as a whole. Since the changes which occur in the universe are subjects of scientific study, we must define precisely what we mean by effort and devise ways of measuring or calculating it accurately.

The technical term for the effort required to bring about a change in our physical environment is *work*. Before we define it however, let us consider just what we mean by change. Careful observation has shown that all change is ultimately a matter of motion. Often the motion is apparent, as when the sun rises, snow falls, or stones are put together to form a building. Sometimes the motion is not so obvious. What, for example, moves when water turns to ice or a piece of metal becomes red hot? Scientific investigation has shown that such phenomena result from the movement and/or the rearrangement of the molecules. When water freezes, the randomly wandering molecules move into an orderly crystalline array where they remain fixed except for small oscillations about their centers of equilibrium; when a solid is heated, its molecules vibrate more and more strongly, and when they are sufficiently agitated, they radiate visible light.

Whenever a stationary object is to be moved (be it an atom

Work and Energy

or a grand piano) or whenever one wishes to change the speed or direction of a moving object, force must be applied. Obviously, the larger the body concerned or the greater the desired change of motion, the stronger the force must be. In other words, work increases with the applied force. That is:

$$\text{Work} \propto \text{force}$$

Work must also depend upon how far the object is displaced from its original position; clearly, it takes twice as much effort to climb two flights of stairs as one. Hence, work must be directly proportional to the distance moved:

$$\text{Work} \propto \text{distance}$$

Both of these relations are simultaneously satisfied by the equation:

$$\text{Work} = \text{force} \times \text{distance}$$

or in symbols:

$$W = f \times d$$

One must apply this definition with caution. It often happens that the object is not free to move in the direction of the force but is constrained to move at an angle to it. Suppose you are pulling a sled as shown in Figure 1. The sled must move horizontally forward, while the force is directed at an upward angle along the rope. Whenever a situation of this sort occurs, the force must be resolved into two components; one parallel to the

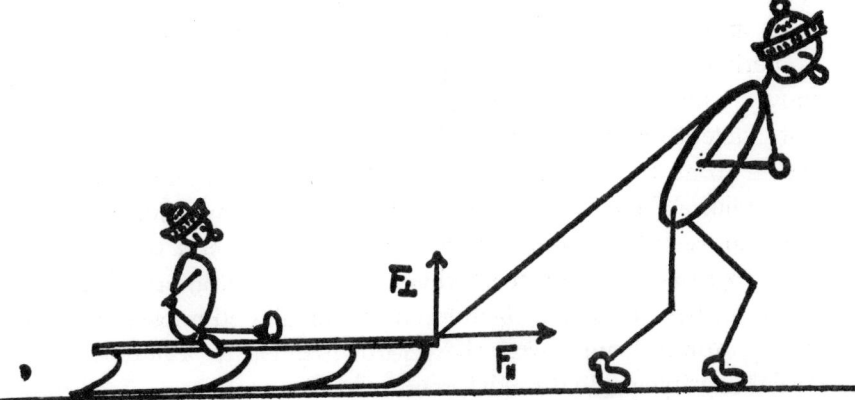

Figure 1.

direction of motion and the other perpendicular to it. These are respectively designated as $F_{\parallel}$ and $F_\perp$ in the figure. Since only the parallel component accomplishes anything, the work done is

$$W = F_{\parallel} \times d$$

Precisely defined, work is the distance moved times that component of the force parallel to the direction of the motion.

Of special interest is the case where the force is perpendicular to the motion, as for example, when a stone is being whirled around the head on a string as shown in Figure 2. Since there

Figure 2.

is no component of force parallel to the motion, no work is done. In fact, if it were not for air resistance, the stone—once set in motion—would move forever without additional effort. In this case the effect of the force is to change the path of the moving object from a straight line to a circle.

In this way gravitational forces hold the planets in (nearly) circular orbits around the sun without doing work; likewise the

Work and Energy

electrostatic force between the atomic nucleus and the electrons has the same effect.

WORK UNITS

The number of possible work units is extremely large since any unit of length (e.g. the inch, foot, yard, centimeter, meter, etc.) can be multiplied by any unit of force (e.g. the pound, dyne, newton, etc.). Altogether, one can easily come up with a hundred or more combinations. As such an abundance is clearly more confusing than useful, scientists and engineers have, for the most part, settled on just four, the *foot-pound*, the *erg*, the *joule*, and the *electron volt*.

The *foot-pound* is the work done when a one-pound weight is lifted one foot. Likewise the *erg* is the work done when a one-dyne weight is lifted a centimeter, and the *joule* is the work done when a one-newton weight is lifted one meter.

The erg, although it is the original metric unit of work, is much too small for most jobs of human interest. It is about the amount of work that a fly would do in crawling a fraction of an inch up a wall. The joule, which is 10^7 times larger, is much more practical. The relation between the joule and the foot-pound is:

$$1 \text{ joule} = 0.7376 \text{ ft-lb}$$
$$1 \text{ ft-lb} = 1.356 \text{ joule}$$

With an error of less than 2%, one can say that the joule is three fourths of a foot-pound.

The erg, joule, and the foot-pound are all excessively large for atomic and nuclear physicists. For their purposes the *electron volt* is much more convenient.

When used for research, elementary charged particles such as electrons, protons, and the like are usually set in motion by placing them in an electrostatic field between two electrodes. See Figure 3. In this figure P_n and P_p are the two accelerating electrodes respectively connected to the negative and positive terminals of the voltage supply B. If an electron e is placed near the negative electrode, a force due partly to the repulsion of P_n and partly to the attraction of P_p will drive that electron

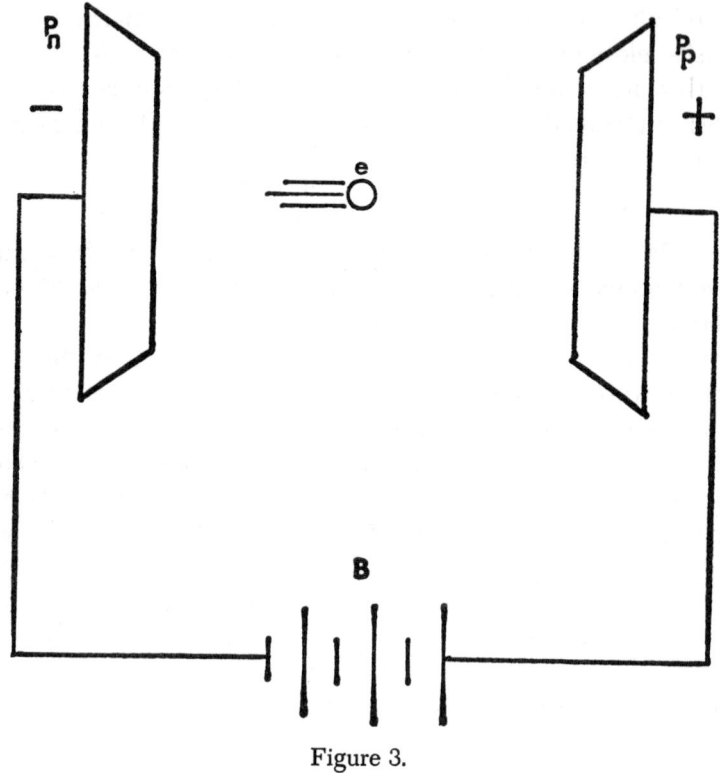

Figure 3.

toward the latter electrode. If unrestrained, the electron will move faster and faster toward the positive electrode until halted by collision. The total work done on the electron is of course the force upon it times the distance traveled.

It can be shown that if the voltage between the two electrodes is fixed at a certain value, the total work done on the electron is the same regardless of the distance between them. The reason is that if the distance is large, the force is comparatively weak; if the distance is decreased, the forces increase in inverse proportion.

Let a particle having a unit charge (1.602×10^{-19} coulomb) be placed between two electrodes so that it will move the entire distance between them. That is to say, the particle almost touches the electrode of the same electrical sign as itself initially. Then, if the potential difference between the electrodes is one volt, the total work done in moving the particle to the opposite electrode is, by definition, *one electron volt.*

If the charge on the particle or the voltage between the electrodes is changed, the amount of work done will change in direct proportion, that is:

$$W = n \times V$$

Where W is the work done in electron volts, n is the number of elementary charges on the particle, and V is the potential difference.

The relationships between the electron volt and the other units of work are not simple. By definition, the potential difference between two points is one *volt* if the electrostatic forces must do one joule of work to move one coulomb of charge from one point to the other. Since the charge on a single electron is 1.602×10^{-19} coulomb, it follows that:

$$1 \text{ eV} = 1.602 \times 10^{-19} \text{ joule or } 1.602 \times 10^{-12} \text{ erg}$$

EXAMPLES

Q. How much work is required to pump 100 gallons of water from a well 75 feet deep?

A. Since a gallon of water weighs 8.32 lb, the total weight to be lifted is 832 lb. If the entire mass were raised at once, a force of this magnitude would have to be applied and the work done would be:

$$75 \text{ ft} \times 832 \text{ lb} = 62{,}400 \text{ ft-lb}$$

If the water is raised a little at a time, the calculation would be different, but the total would remain the same.

Q. A man weighing 150 lb climbs a flight of stairs. If the distance between floors is ten feet and the stairs rise at an angle of 45°, how much work must the man do?

A. The total distance the man moves is 14.14 feet. However, it is not correct to multiply this distance by the man's weight, since the force and the direction of movement are not parallel. According to what was said earlier, one would have to multiply the distance by the component of force parallel to it; however it is equally correct to multipy the total force by the net displacement parallel to it instead. In this problem, the latter procedure is far more convenient. The force of gravity is vertical and the vertical displacement is ten feet. Hence the work done is:

$$10 \text{ ft} \times 150 \text{ lb} = 1{,}500 \text{ ft-lb}$$

Q. An automobile weighing two tons travels one mile on a road that rises ten feet per mile. If the road friction and the wind resistance add up to an opposing force of 500 lb, how much work is done?

A. In this example, two things are accomplished. First, the car is lifted a total of ten feet requiring:

$$\text{Work of lifting} = 10 \text{ ft} \times 4{,}000 \text{ lb} = 40{,}000 \text{ ft-lb}$$

In addition, the car is pushed one mile against a force of 500 lb. The work done is

$$5{,}280 \text{ ft} \times 500 \text{ lb} = 2{,}640{,}000 \text{ ft-lb}$$

The total work is therefore:

$$40{,}000 + 2{,}640{,}000 = 2{,}680{,}000 \text{ ft-lb}$$

ENERGY

Energy, perhaps the most important concept in all science, is closely related to work. Unfortunately, it is difficult to give a definition that is rigorous yet meaningful to the beginner. The following statement is probably as good as any:

The energy content of a system is equal to the amount of work that the system can do to accomplish a specified task.

For example, consider a hydroelectric station driven by a mountain lake situated 100 feet above the turbines as shown in Figure 4. If the lake has an area of five square miles and an average depth of ten feet, how much work can the system do? In this example the word system includes the dynamos, turbines, all auxiliary equipment, and of course the lake itself.

If the entire lake were emptied through the turbines, the

Work and Energy

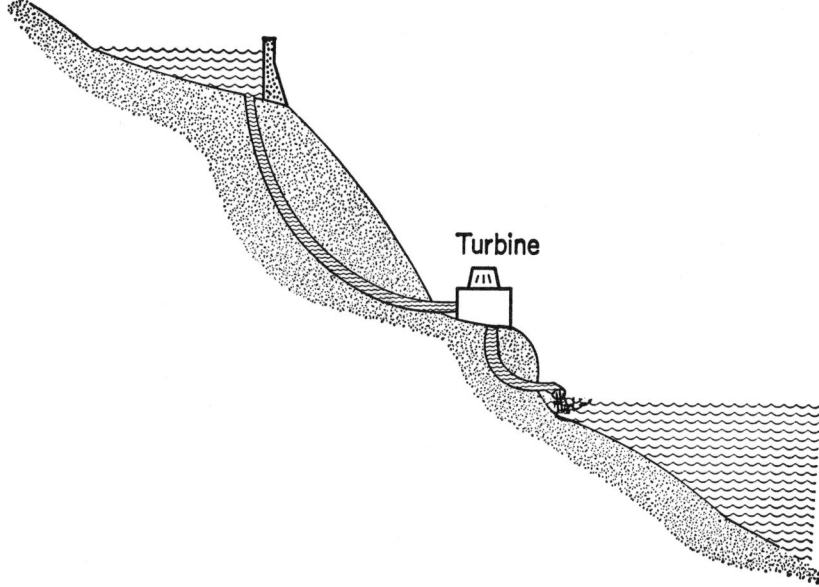

Figure 4.

effect would be the same as if the total quantity of water had been lowered 100 feet. The work done would therefore be the weight of the water times the distance moved. Since a mile is equal to 5,280 ft and the density of water is 62.4 lb/ft³, the work done would be:

$$100 \times 5 \times 5{,}280^2 \times 10 \times 62.4 = 8.70 \times 10^{12} \text{ ft-lb}$$

Hence we say that the system contains $+ 8.70 \times 10^{12}$ ft-lb of energy. The plus sign indicates that the work can be done *by* the system *for* us. Notice that energy has the same units of measure as work.

It is very important to recognize that the energy content of a system varies according to the job to be done. Consider again the lake just mentioned. Instead of using the water to generate electricity, suppose that we want to use it to irrigate some fields fifty feet higher up the mountain. Since the water will not flow up of its own accord, we must bring in some outside machinery to lift it. The total amount of work required is equal to the weight of the water times the distance it is to be raised, namely:

$$50 \times (5 \times 5{,}280^2 \times 10 \times 62.4) = 4.35 \times 10^{12} \text{ ft-lb}$$

For the purpose intended, we say that the energy content of the system is -4.35×10^{12} ft-lb, the minus sign indicating that the work must be done *by* us *on* the system.

The concept of positive and negative energy values is extremely important, especially in atomic and nuclear physics. Consider, for example, the nucleus of Ca-41. It has one too many protons for stability and therefore must decay sooner or later to K-41. Two processes for transforming a proton to a neutron exist, electron capture and positron emission. The latter requires about one million electron volts more than the first.

The Ca-41 nucleus has sufficient energy for the capture process but not for positron emission. Hence the energy of the system is either positive or negative, according to which reaction one has in mind.

KINETIC ENERGY

Kinetic energy is the energy that an object has by virtue of its motion. A rifle bullet offers an excellent example. At rest, it is a harmless piece of metal; moving with a speed of several hundred feet per second it can bore a hole through a thick sheet of metal, a task requiring a great deal of effort. In doing work, a moving body must of course slow down, coming to rest when its supply of kinetic energy is exhausted.

For bodies moving slowly compared to the speed of light, the relation between kinetic energy and speed is given with sufficient accuracy by the formula:

$$\text{K.E.} = \tfrac{1}{2} mv^2$$

where m is the mass of the body and v its speed. This formula requires careful attention to units. The following sets are compatible:

gram,centimeter/second,erg
kilogram,meter/second,joule
slug,foot/second,foot-pound

The *slug* is a unit of mass used by American engineers equal to 14.59 kg. It is also approximately the mass of an object weighing 32 lb.

Work and Energy

For objects moving faster than, say, a tenth of the speed of light, the preceding formula is too inaccurate. For them one must use the relativisitically correct expression:

$$\text{K.E.} = mc^2 \left\{ \frac{1}{\sqrt{1 - \frac{v^2}{c^2}}} - 1 \right\}$$

where c is the velocity of light in vacuum, namely 3.00×10^8 m/sec. When v is much less than c, this formula and the simpler one given earlier yield very nearly the same answers.

POTENTIAL ENERGY

Any form of energy except kinetic is called *potential*. Potential energy can be further subdivided into a number of kinds, the more important being:

Mechanical (raised weights and coiled springs)
Chemical (batteries and fuel burning engines)
Nuclear (reactors and bombs)
Electrical (charged capacitors)
Magnetic (magnets and transformers)
Heat (thermal engines)

In every case, the quantity of energy in a system is determined by measuring or calculating the amount of work that it can do (if its energy content is positive) or has to be done on it (if its energy content is negative).

THE CONSERVATION OF ENERGY

Energy can neither be created nor destroyed. One form can however be transformed into another. Since it often happens that energy, initially available, is converted to a form that cannot be utilized or even detected by reasonably practical means, the conservation law is by no means self-evident. It was, in fact, discovered long after a number of other major physical laws had become thoroughly familiar.

As a simple example, consider a weight suspended at a considerable height and then dropped to the ground. For simplicity we will ignore the slight effects of air resistance. While it is hanging, the weight possesses potential energy since it is capable of doing work, such as, for example, driving a clock mechanism. As it falls, it loses potential energy but progressively picks up speed and simultaneously kinetic energy according to the relation K.E. $= \frac{1}{2}mv^2$. It can be shown that in its downward motion, the potential energy lost is just equal to the kinetic energy gained at each point. Just before striking the ground, all the potential energy has disappeared and the kinetic energy has reached its maximum value. At the moment of impact, the gross motion of the object is halted, being converted into increased agitation of the molecules both of the object and the ground struck. This additional molecular motion constitutes an increase in heat content and manifests itself through a rise in temperature. Soon however, this heat is dissipated throughout the environment and is, for any practical purpose, permanently lost.

PROBLEMS

1. A ten ton weight is suspended 75 feet above the earth. How much potential energy does it possess? (Hint: How much work would be required to lift it up there in the first place?)

2. Suppose the weight in problem *one* were to fall. (a) How much kinetic energy would it have just before striking the ground? (b) What would be its speed at that moment?

3. A gasoline engine is used to pump water from a well 250 feet deep. When burned, a gallon of gasoline yields 10^8 ft-lb of work, but the engine is only 20% efficient. If the well is to deliver a thousand gallons a minute 24 hours a day, how much gasoline must be supplied per week? A gallon of water weighs 8.32 lb.

4. Consider three bodies whose respective speeds are 0.01, 0.1, and 0.9 of the speed of light. Calculate the kinetic energy of each using both the low-velocity approximation and the relativistically correct formula. What is the percent error for each of the three bodies?

Chapter III UNITS AND CONSTANTS OF MICROSCOPIC PHYSICS

BY THE TERM *Microscopic Physics* we mean: molecular physics, atomic physics, nuclear physics, and particle physics. The last mentioned includes not only the study of material particles such as electrons, protons, mesons, and the like, but photons as well.

Because atoms and molecules are so small, the familiar units of measure (such as the meter and the kilogram) are usually too inconvenient for routine use. Consequently several special units have been devised.

Most physical quantities are combinations of length, mass, time, and/or electrical current. For example, *velocity* is length divided by time; *density* is mass divided by volume (that is, length cubed); and so on. Considerably more complicated combinations are possible; *force* is mass times length divided by time squared.

For each of the four named quantities an internationally recognized unit of measure has been defined:

Length .... meter
Mass .... kilogram
Time .... second
Current .... ampere or coulomb/second

Exactly how these units were established and are maintained is a fascinating but long and difficult story which we must forego in this book.

TIME: Units of time appropriate to particle physics are obtained using the standard metric prefixes:

milliseconds (msec) ....10^{-3} sec
microsecond (μsec) ....10^{-6} sec
nanosecond (nsec) ....10^{-9} sec

Some nuclear processes such as alpha and beta decay require very long intervals for which one uses minutes, hours, days, or years as appropriate. Certain fundamental particle reactions may take as little as 10^{-22} sec. For such extremely short intervals, one often uses the second together with *ten* raised to the necessary negative power.

LENGTH: Atomic units of length are more numerous and more complex than those of time. Two metric subdivisions below the millimeter are used, 10^{-6} and 10^{-9} meter. The former is not called the *micrometer* as one would expect according to the rule, since that word happens to be the name of a certain measuring instrument. Instead, the word *micron* is used. The latter unit, 10^{-9} m, has usually been called the *millimicron* instead of the *nanometer* although the latter word is gaining favor.

For many years physicists used a unit called the *angstrom* which is 10^{-10} m or 0.1 nm. While this unit appears to be going out of fashion, it pervades all the older literature. The loss of this unit is to be regretted since in most solid substances, the spacing between adjacent atoms is usually about one angstrom.

Still another unit, the XU (for x-ray unit) is very nearly—but not exactly—equal to 10^{-13} m or 10^{-7} micron. It has fallen into disuse, for which all may be grateful.

MASS: While the masses of atomic and nuclear particles can be stated in grams or metric subdivisions thereof, it is exceedingly awkward to do so. Two special units are in current use, the *electron mass* (m) and the *atomic mass unit* (AMU). The former is of course the rest mass of the electron. The latter was until recently set equal to $\frac{1}{16}$ of the mass of the oxygen-16 atom. A few years ago the definition was altered to $\frac{1}{12}$ of the carbon-12 atom. This change barely altered the size of the AMU but it did result in a number of practical simplifications. The relationships between the electronic mass and the atomic mass unit and the kilogram are:

$$1 \text{ m} = 9.109 \times 10^{-31} \text{ kg}$$
$$1 \text{ AMU} = 1.6603 \times 10^{-27} \text{ kg}$$

The mass of any atom in AMU is very nearly equal to its mass number. For example:

Nuclide	Mass of one atom (AMU)
H-1	1.007825
C-12	12.0000....
O-16	15.9949
Ra-226	226.0254

The masses of the neutron and the proton are each less than one per cent greater than one AMU.

CHARGE: The standard unit of electrical charge (the *coulomb* or the *ampere-second*) is much too large for particle physics. Since the quantity of electricity on every charged particle is exactly equal to that of the electron (or an integral multiple thereof) that quantity, called the *electronic charge* is universally used as a unit. Its symbol is e. Its value is 1.602×10^{-19} coulomb.

WORK AND ENERGY: At the molecular and submolecular level work and energy are nearly always specified in *electron volts* or other units derived by using the metric prefixes:

kilo-electron-volt (keV) .... 10^3 eV
mega-electron-volt (MeV) .... 10^6 eV
giga-electron-volt (GeV or BeV) .... 10^9 eV

FORCE: Although phenomena are initiated by forces at the atomic and subatomic levels as elsewhere in nature, one is seldom interested in their magnitudes. Consequently one has no need for special force units.

CONSTANTS

Several physical constants are of great importance in microscopic physics. Two have been mentioned, the charge and the mass of the electron. Others are:

THE VELOCITY OF LIGHT: This quantity, usually represented by the letter c, is equal to 2.998×10^8 m/sec. All photons move at this speed in vacuum. The speed of an electron is also near this value whenever its kinetic energy exceeds a keV or so. Heavier particles require correspondingly higher energies for velocities in this range. Whenever it is necessary to state the speed of a subatomic particle, one often uses the quantity β which is defined as the ratio of the speed of the particle to that of light, that is:

$$\beta = \frac{v}{c}$$

According to the theory of relativity, no entity bearing mass, energy, or information can travel faster than the velocity of light.

AVOGADRO'S NUMBER: If one has a quantity of any element such that the mass in grams is numerically equal to its atomic weight, then one is said to have one gram-atom of the element in question. Thus 1.0078 g of hydrogen, 26.98 g of aluminum or 197.1 g of gold would in each case be one gram-atom.

A basic fact of chemistry is: The number of atoms in one gram-atom is constant for all elements, namely 6.023×10^{23}. This quantity is known as *Avogadro's Number*. It must be used, directly or indirectly, whenever one wishes to calculate the specific activity of any radioactive material.

Most elements are isotopic mixtures, and the atomic weights given in chemical tables are averages over the various isotopes present. Avogadro's Number is also useful when dealing with individual nuclides. If the mass of a particular sample of a certain nuclide is equal to its nuclidic weight, then one is said to have a gram-atom of that nuclide, and the number of atoms present is again equal to 6.023×10^{23}.

PLANCK'S CONSTANT: This important quantity (symbol h) occurs in many situations. For our purposes it is the ratio of the energy of a photon to its frequency:

$$h = \frac{E}{f} = 6.626 \times 10^{-27} \text{ erg-sec}$$

$$= 4.135 \times 10^{-15} \text{ eV-sec}$$

The fact that h is constant means that photons of high frequency (and short wavelength) such as x-ray and gamma photons have a great deal of energy, while those of low frequency (and long wavelength) such as infrared and radio photons have but little energy.

PROBLEMS

1. What is the mass of one trillion atoms of uranium? The atomic weight of uranium is 238.03.

2. Photons of 5,500 angstroms wavelength lie in the middle of the visible spectrum. What is the energy of one such photon (a) in joules and (b) in electron volts? Note: The frequency of a photon is equal to the velocity of light divided by the wavelength.

3. A gamma ray photon has an energy of one MeV. What is its wavelength?

4. A certain radio station broadcasts on a frequency of 1,000 kilocycles per second with a radiated power of one kilowatt. What is the wavelength of the emitted photons? What is the energy of each? How many photons are emitted per second?

Chapter IV FUNDAMENTAL PARTICLES

BACK IN THE DAYS when the electron, proton, and neutron were the only subatomic particles known, they were called fundamental particles in the belief that the philosophical atoms had at last been found. Since that time physicists have discovered more and more "fundamental" particles, several dozen having now been identified. For elementary understanding of applied nuclear physics however, only the following are important:

proton
neutron
electron (or negatron)
positron (or anti-electron)
neutrino
anti-neutrino

Of these only the first three ordinarily exist within the atom. The others may appear in nuclear reactions.

Three important physical properties of these particles are their mass, electrical charge, and diameter. Their values are listed in Table I. Besides being extremely small by usual standards, these quantities have several interesting and important features.

Concerning the charged particles, the quantity of electricity on each is the same except for a difference in sign. Since no smaller electrical charge has ever been found, and since all larger aggregates of charge appear to be integral multiples of this value, we now believe that charge is indeed granular, the size of the individual grain being $\pm 1.602 \times 10^{-19}$ coulomb.

Fundamental Particles

TABLE I

Particle	Charge (Coulombs)	Mass (Kilograms)	Mass (MeV)	Radius (Meters)
Proton	$+1.602 \times 10^{-19}$	1.6603×10^{-27}	938.256	1.6×10^{-15}
Neutron	zero	1.67482×10^{-27}	939.550	1.6×10^{-15}
Electron	-1.602×10^{-19}	9.109×10^{-31}	0.511006	4.5×10^{-15}
Positron	$+1.602 \times 10^{-19}$	9.109×10^{-31}	0.511006	4.5×10^{-15}
neutrino	zero	zero	zero	?
anti-neutrino	zero	zero	zero	?

Contrary to expectation, no simple relationships exist among the particle masses. Those of the proton and the neutron are nearly equal, although that of the neutron is slightly greater. The electron and the positron have the same mass within the limits of measurement, a value about 1,836 times smaller than that of the proton. The mass of the neutrino and the anti-neutrino is at least a thousand times less than that of the electron and is thought to be exactly zero.

The notion of radius cannot be rigorously applied to these particles, since they are not hard objects with sharp, rigid boundaries like a steel ball or a marble. Instead, there is a transition zone of indefinite extent from the interior of the particle to the free space outside. Conventionally the radius is defined as the distance from the center of the particle to the middle of the transition zone. It is possibly surprising that the electron is about as large as the much more massive neutron and proton.

Apart from the difference in electrical charge, the proton and the neutron look suspiciously alike. If one considers other properties not discussed here (e.g., spin, baryonic number, strangeness, and others) they look even more alike. It is now generally accepted that the neutron and the proton are really different manifestations of the same particle. For this reason the name *nucleon* is applied to both. In nuclear physics, one often says *nucleon* when stating a fact equally true of either a neutron or a proton just as one says *person* when making a remark applicable to both men and women.

TRANSFORMATIONS OF ELEMENTARY PARTICLES

THE PROTON: Although the proton has never been resolved into units more elementary than itself, it may undergo certain transformations, two of which are most important. For example, it may combine with an electron to produce a neutron and a neutrino. The reaction is expressed by the equation:

$$p + e \rightarrow n + \nu$$

This reaction is endothermic in free space, meaning that it can take place only if energy is added from without. Hence there is no danger that the enormous quantity of hydrogen in the ocean (each of whose atoms consists of a proton and an electron) will be transformed into a neutron sea, because the energy needed for the reaction is not there. Likewise the water of the ocean cannot dissociate into hydrogen and oxygen gas because that reaction is also endothermic.

While the transformation of a proton and an electron into a neutron and a neutrino cannot occur in free space, it can and does occur in certain atomic nuclei. One can explain this fact either by saying that within such nuclei, the necessary energy for the reaction is available in useful form, or by saying that within such environments, the reaction becomes exothermic. Both statements are true, and in fact equivalent, but represent slightly different points of view. The proton may also decay directly without the assistance of another particle into a neutron, positron and a neutrino:

$$p \rightarrow n + e^+ + \nu$$

This reaction is even more endothermic than the preceding and therefore occurs only in those nuclei with an ample supply of energy. Nuclei in which either of these two processes may occur are called "proton rich." Their decay produces a new, more nearly stable, nucleus having one more neutron and one less proton than the original.

THE NEUTRON: The neutron may decay into a proton, an electron, and an anti-neutrino:

$$n \rightarrow p + e + \tilde{\nu}$$

This reaction is exothermic in free space and therefore occurs spontaneously. Because of this reaction, a free neutron will survive only about fifteen minutes on the average. Within the atomic nucleus the situation is usually different; in some, the reaction is still exothermic so that the nucleus undergoes nuclear disintegration. However the average life of the neutron may be anything from a fraction of a second to billions of years for the different types of nuclei. In others, the reaction is endothermic with the result that the neutrons present last indefinitely. Those in which neutron decay occurs spontaneously are called "neuton rich." The reaction transforms such a nucleus into another having one more proton and one less neutron than the original. It is important to remember that the total number of nucleons remains unchanged.

The result of these two processes, one which increases the neutron/proton ratio and the other which decreases it, is to severely limit the number of possible combinations of neutrons and protons that can form stable nuclei. In fact, for a given total number of nucleons, there is usually only one or two combinations. For the lighter elements, the number of protons Z in the nucleus must be nearly equal to the number of neutrons N for stability. In heavier nuclei, the N/Z ratio must be somewhat greater than one for reasons to be explained later.

ELECTRONS AND POSITRONS: Electrons and positrons are both stable particles—as long as they avoid each other. But whenever one lingers in the neighborhood of the other, as is likely to happen since they have opposite charges, both vanish completely. In their stead appear two photons traveling in opposite directions each with an energy of 511 keV:

$$e^+ + e^- \rightarrow 2\gamma$$

Since matter on earth contains countless billions of electrons, no positron can exist for more than a fraction of a second.

Another reaction one would expect to occur is:

$$n + e^+ \rightarrow p + \bar{\nu}$$

However this reaction is not observed because the positron is always destroyed by another electron long before it can find a neutron to react with.

THE CONSERVATION LAWS

Most of the basic facts of nature can be expressed as *conservation laws*, that is, statements saying that certain quantities cannot change in value. The particle reactions just discussed illustrate several of these.

THE CONSERVATION OF ELECTRICAL CHARGE: According to this law the total amount of electrical charge in any system cannot change. Thus in the reaction:

$$p + e \rightarrow n + \nu$$

the total initial charge is zero since the charges of the proton and the electron are equal and opposite. Since the two resulting particles are both electrically neutral, the final value of the total charge is also zero as the law requires. The other reactions also conserve charge as you may verify.

THE CONSERVATION OF LEPTONIC NUMBER: The very lightest particles in nature, the electron, positron, neutrino, and anti-neutrino, are called *leptons*, from the Greek word λεπτός meaning *light in weight*. By *leptonic number* we mean the number of leptons present in any given system. Thus the neutral helium atom, which contains two electrons, has a leptonic number of 2 and so on.

Nature seems to require that every particle have a corresponding anti-particle, that is to say, one whose physical parameters are equal and opposite to those of the real particle. The electron and positron (or anti-electron) constitute one such pair. The electron has a negative charge and its anti-particle a positive charge. The leptonic numbers are also reversed. If we count the electron as +1 lepton, then we must count the positron as −1 lepton. The same applies to real and anti-neutrinos. In the reaction:

$$p + e \rightarrow n + \nu$$

we begin with +1 lepton (the electron) and end with +1 lepton (the neutrino). In the reaction:

$$p \rightarrow n + e^+ + \nu$$

we begin with no leptons at all and end up with $-1 + 1 = 0$ leptons, since the positron must be counted as −1 and the neutrino as +1. In all physical reactions so far discovered, the leptonic number is conserved.

Conservation of Baryonic Number: The proton, neutron, and all heavier *fundamental* particles are called baryons from the Greek word βαρύς meaning *heavy*. The rules of baryonic number are the same as those for leptonic number. However, in the reactions discussed here no anti-baryons occur. Antiprotons and antineutrons arise only in reactions demanding far more input energy than is available in an atomic nucleus. Hence for our purposes the conservation of baryonic number is especially simple. In the reaction:

$$n \rightarrow p + e + \tilde{\nu}$$

we begin with +1 baryon (the neutron) and end with +1 baryon (the proton). In the reaction:

$$e^+ + e^- \rightarrow 2\gamma$$

there are no baryons either before or after.

The Conservation of Energy: As elsewhere in nature, so in nuclear reactions, the conservation of energy is one of the most important characteristics. According to this principle, the total energy of the system before any reaction occurs is equal to the total energy afterward.

As far as the fundamental particles are concerned, energy may exist as:

1. Rest mass energy of the particle according to the equation $E = mc^2$.

2. Kinetic energy.

3. Electromagnetic energy in the form of photons.

To see how the conservation law applies in practice, consider the reaction:

$$n \rightarrow p + e + \tilde{\nu}$$

Before the disintegration occurs, the neutron is probably at rest or moving very slowly so that its kinetic energy is negligible. Hence the only energy in the system is the rest mass energy of the neutron. After the reaction there are three particles all in motion. Therefore the energy could presumably be divided into six parts:

1. Rest mass energy of the proton E_p
2. Rest mass energy of the electron E_e
3. Rest mass energy of the antineutrino E
4. Kinetic energy of the proton KE_p
5. Kinetic energy of the electron KE_e
6. Kinetic energy of the anti-neutrino $KE_{\bar{\nu}}$

The situation is in fact a little simpler. Since the neutrino has no rest mass, its rest mass energy is obviously zero. Furthermore, in this reaction the recoil velocity of the proton will be very slow and hence its kinetic energy will be negligibly small for most purposes. With these simplifications, the energy balance equation can be written:

$$E_n = E_p + E_e + KE_e + KE_{\bar{\nu}}$$

We now see why the reaction:

$$p \rightarrow n + e^+ + \bar{\nu}$$

cannot occur in free space. The only energy available to begin with is the rest-mass energy E_p of the proton. Since this is less than the rest mass energy E_n of the neutron, the proton does not even have enough energy to create the neutron to say nothing of the other two particles.

In the reaction:

$$e^+ + e \rightarrow 2\gamma$$

the available energy is the rest mass energy of the electron and the positron. Little kinetic energy is present since the reaction does not occur until the two particles are nearly at rest. The energy balance equation is therefore:

$$E_e + E_{e^+} = 2E_\gamma = 1.02 \text{ MeV}$$

PROBLEMS

1. A neutron at rest in empty space disintegrates. What is the total kinetic energy of the decay products?

2. How much energy must the nucleus supply to trigger the reaction:
$$p + e \rightarrow n + \nu ?$$
3. How much energy must a nucleus supply to trigger the reaction:
$$p \rightarrow n + \tilde{e} + \nu ?$$

Chapter V ELEMENTS, NUCLIDES, ETC.

FUNDAMENTAL TO PHYSICS and chemistry are the words *element, nuclide, isotope, isobar,* and *isotone.* As they are often misused and misunderstood, let us take a little time to define them with especial care.

First of all, we must clear away a couple of common misconceptions. Although statements found in certain textbooks suggest otherwise, none of these words denotes a single atom or nucleus but rather bulk matter. It is perfectly possible to have a cupful of an element or a pound of a nuclide. Neither do any of these words imply radioactivity. It is quite wrong to say *isotope* when you mean *radioactive material.* Very few naturally occurring isotopes are radioactive—fortunately.

ELEMENT: The word *element* means a quantity of matter each of whose atoms contains the same number of protons. This number is called the *atomic number* of the element and is represented by the letter Z. Commonplace articles composed of a single element include a gold ring, its diamond setting, a piece of copper wire, and an aluminum pan.

The atomic numbers of the above mentioned elements are:

    Gold 79
    Diamond (crystalized carbon) 6
    Copper 29
    Aluminum 13

Years ago chemists devised symbols to represent the various elements. For the more common ones, the symbol is a single capital letter such as C for carbon and H for hydrogen. For most elements the symbol consists of two letters written together, the second being lower case. Examples are *La* for lanthanum, and *Ne* for neon. Most symbols are obviously related to the English name of the element. A few are derived from Latin as *Au* for gold (aurum), *Fe* for iron (ferrum), and *Ag* for silver (argentum). *W* for tungsten comes from the German Wolfram.

How many elements are there? Until recently many chemists would have answered with assurance, "Exactly 92." Now the situation is not so simple. During the two centuries before World War II many elements were discovered with atomic numbers ranging from *one* (hydrogen) to 92 (uranium). Since uranium is fairly common and no trace whatever was found of any element beyond, it was plausibly assumed that some unknown natural law prevented the existence of heavier elements. It was also plausibly—but wrongly—assumed that samples of all elements of lower atomic number would be found upon the earth.

Gaps remained in the periodic table however. There were several between lead (82) and uranium, and curiously, positions 43 and 61 remained empty. Recurring reports of discovery proved false. We now know why these gaps are there. The elements involved are unstable and decay so rapidly that the original supply—if any—has long since disappeared.

Since the discovery of controlled atomic fission, men have created many elements hitherto unknown. Of the undiscovered elements mentioned above, samples have now been made of each and also of every element beyond uranium up to 106 as of the time of writing. There is good reason to believe that elements of still higher atomic number can be prepared. How high it will be possible to go is not known. The limiting factors are probably human: interest, patience, and money.

To return to the original question: How many elements are there? Eighty-odd occur naturally on earth; the exact number is in dispute—or would be—if anyone really cared. Small quantities of at least twenty more have been prepared by man. In the interiors of the stars, where nuclear transformations are in progress, still more no doubt exist. If one could assay the entire universe, perhaps as many as 200 different elements could be identified.

NUCLIDES: The word nuclide is more restrictive than the word *element*. To be classified as a nuclide, a substance must not only have the same number of protons in every atom but the same number of neutrons as well. Of course the number of protons per atom need not and usually does not equal the number of neutrons. The number of neutrons in an atomic nucleus is called the *neutron number* and is represented by the letter N.

For most purposes the value of the neutron number is not too useful. Of considerably more importance is the *mass number* A, which is the total number of particles, neutrons plus protons, in the nucleus, that is:

$$A = N + Z$$

Symbols for the various nuclides are formed by adding one to three sub- or superscripts to the symbol for the chemical element. Of these the first specifies A, the second Z and the third N. For example, naturally occurring aluminum, which is not only an element, but also a nuclide, may be symbolized:

$$\text{Al-27} \quad \text{or} \quad \text{Al}^{27} \quad \text{or} \quad ^{27}\text{Al}$$

The first form, Al-27, is used in typing or printing when superscripts are not available. The second form, Al^{27}, is to be found in most older books and scientific articles, but is no longer approved. The currently preferred form is ^{27}Al.

Although the number of protons in a nucleus is unambiguously implied by the elemental symbol, most people do not remember off hand the atomic numbers of any but the commonest elements. As an aid to the reader, Z may be expressed as a left subscript:

$$_{13}\text{Al}^{27} \quad \text{or better:} \quad ^{27}_{13}\text{Al}$$

A number of naturally occurring elements are also nuclides especially if Z is odd. Most elements, however, are nuclidic mixtures, although frequently one nuclide is overwhelmingly predominant. Carbon, for example, consists of about 99% C-12, about 1% C-13, and a trace of radioactive C-14.

Occasionally one may want to explicitly specify N as well:

$${}^{27}_{13}\text{Al}_{14} \quad \text{or better:} \quad {}^{27}_{13}\text{Al}_{14}$$

where N is shown by the right subscript.

Since the chemical properties of atoms are determined by the electrostatic attraction between the protons in the nucleus and the orbital electrons, it follows that all the nuclides of a particular element have the same chemical properties. There are minor deviations, but they are insignificant except in the case of hydrogen. Since each atom of deuterium ^{2}H is twice as heavy as an atom of ordinary hydrogen ^{1}H, there are appreciable differences in reaction rates and molecular stability. In fact, some living organisms cannot survive if their body water is replaced with deuterium oxide, that is, water prepared from ^{2}H instead of ^{1}H.

ISOTOPES: *Isotope* is one of the most misused of all scientific terms. The usual error is that most people say *isotope* when they mean *nuclide*.

Perhaps the difference between the two words is best illustrated by considering the words *man* and *brother*. Both mean a male human being; the word *brother* however implies a relationship between the person in question and some one else. Exactly the same can be said of the words *nuclide* and *isotope*. Both denote a substance that has the same values of N and Z in each of its atoms. *Nuclide* implies nothing more than this; *isotope* indicates membership in a family. Specifically:

All nuclides having the same value of Z are isotopes each of the others.

Thus ^{1}H, ^{2}H, and ^{3}H are isotopes of hydrogen, ^{12}C, ^{13}C, and ^{14}C are isotopes of carbon and so on.

These two rules will help you use the words *isotope* and *nuclide* correctly:

1. When in doubt, say *nuclide*, not *isotope*.
2. The word *isotope* should usually be followed by *of* and the name of the element concerned. Thus *an isotope of copper*, or the *most abundant isotopes of tin* is proper. The sentence, *Technecium has no stable isotopes*, is correct both scientifically and grammatically.

ISOBARS: Nuclides can be classified according to their mass numbers. For example, ^{13}B, ^{13}C, and ^{13}N all have mass number 13. Such a group is called an *isobaric family* and each member is an *isobar* thereof. This concept is quite important because whenever a nuclide undergoes beta decay, the daughter is always an isobar of the parent.

Most isobaric families have a fair number of known members. Of these, some are proton rich, some are proton poor, and one or two have an optimum balance of protons and neutrons. These last are stable; all the others are radioactive and will decay until only stable forms are left. In the example given above, both ^{13}B and ^{13}N will decay to ^{13}C which is stable.

ISOTONES: A set of nuclides all having the same value of N is called an *isotonic family* and each of its members an *isotone*. Such families are of little practical importance and the word *isotone* is seldom heard.

Chapter VI THE STRUCTURE OF BULK MATTER

For many years it was thought that all bulk matter consists of molecules which are in turn made up of atoms. One still finds statements to that effect in many popular books and even in elementary texts by authors who should know better. Actually a great many natural substances are not so constituted. Let us therefore pause to discuss the commoner types of material structure.

The Noble Gases: The simplest substances found in nature are the noble gases: helium, neon, argon, krypton, xenon, and radon. These consist of isolated atoms and show no evidence of higher structure in the gaseous state. While these elements can be reduced to liquids and even solids at very low temperatures where of course some large-scale order must exist, this behavior is too complex to discuss here and is not relevant to our purpose.

In these gases, no molecules exist unless one insists on calling the single atoms such. However this sort of word juggling is a linguistic exercise, not physics or chemistry.

The Molecular Gases: Most gases, however, do consist of molecules which in turn are made up of either neutral atoms or of ions. The molecules dart about independently in the space available with practically no interactions among them except for collisions such as occur among billiard balls. Unless contained, the rapidly moving molecules will fly further and further apart until they are completely dispersed. Such being true, one may ask why the atmosphere of the earth has not disappeared into outer space. The answer is that the gravitational attraction of the earth is strong enough to prevent that from happening; any molecule shooting spaceward soon falls back of its own weight. In other words, the earth's gravitational field, although it has no material walls, acts as a container. Heavenly bodies much smaller than the earth cannot hold an atmosphere. If Mars or the moon ever had one, it vanished long ago. There are still a few wisps of gas about Mars; the region near the surface of the moon is an exceedingly good vacuum.

If the molecules of a gas (other than the noble gases) are subjected to sufficient chemical or physical stress, they will break apart into either atoms or ions. If the number of electrons in each is equal to the number of protons, the particle is called an *atom;* if the numbers are unequal, an *ion.* Ions are positive or negative according as the protons or electrons predominate. Some molecules such as hydrogen chloride split into ions, others such as oxygen into atoms.

LIQUIDS: As the temperature of a gas is reduced, its molecules move slower and slower. If the pressure is held constant, the gas will also shrink in volume so that the molecules come closer and closer to one another. Eventually they will be near enough for attractive intermolecular forces to become effective. When that happens, the entire mass will stick together of its own accord. Usually (but not always) the resulting substance is a liquid. If so, the molecules can still slide over one another and migrate throughout the volume. Those at the surface may also acquire enough energy to fly off into space. This latter process is called *evaporation.*

Because the intermolecular forces are so weak, a liquid cannot hold a shape of its own but conforms to that of the containing vessel. A blob of liquid in empty space, free from outside influences, will become spherical since those forces pull it into the most compact form possible.

A liquid may contain ions as well as molecules. Even in the purest water, a few of the molecules are dissociated into negative hydroxyl (OH^-) and positive hydrogen (H^+) ions. When dissolved into water, many inorganic acids, bases, and salts dissociate more or less completely into ions. On the other hand organic substances such as sugar, alcohol, and the like separate into individual, intact molecules which migrate among those of the solvent.

AMORPHOUS SOLIDS: If a liquid is cooled, its molecules become progressively less agitated and move still nearer. The forces between them become stronger and eventually the molecules become locked together. The substance is then a rigid mass with a shape of its own, which can be altered only by sufficient force. In some materials such as glass, the molecules occupy about the same positions that they had in the liquid phase. Since they are jumbled together without much regularity, such solids are called *amorphous*. They are literally frozen liquids. For such substances, there is no definite melting point. The transition occurs over a broad range of temperatures.

SINGLE CRYSTALS: Most substances do not solidify into a glass-like state. Instead, once a definite temperature called the freezing point, is reached, molecules, atoms, or ions (depending on the nature of the substance) begin to arrange themselves into a regular array, something like the bricks in a wall. The resulting solid mass is called a crystal. Some substances such as sodium chloride (common table salt) or quartz can form large single crystals. Others form countless tiny crystals that cling tightly together.

There are several different geometrical patterns into which the crystal elements may fall. One of the simplest is called the cubic lattice and is found in many common substances including sodium chloride. As Figure 5 shows, this substance is made up of positive sodium ions and negative chlorine ions in alternation so positioned as to form the corners of a series of stacked cubes.

Crystals such as sodium chloride show no evidence of molecular structure. If one is to retain the molecular concept at all, one must say that the entire crystal is a single enormous molecule.

A perfect crystal would be one in which every position in the lattice is filled by the correct particle. In nature and even in the laboratory, where conditions can be rigorously controlled, perfect crystals of any size are impossible to achieve. Defects are of two kinds; impurities and geometrical errors.

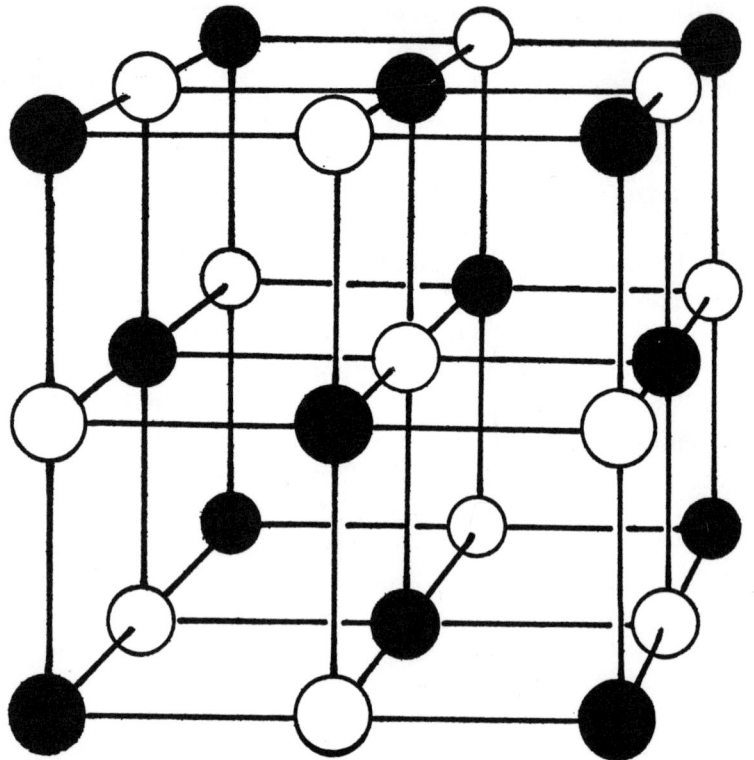

Figure 5.

To visualize defects of the first kind, imagine a red brick wall into which a careless or color-blind mason has scattered a few yellow bricks. Thus, sodium chloride, for example, probably contains a few ions of potassium, fluorine, and iodine scattered through it.

Geometrical defects can also be found in a brick wall: cracks, missing bricks, extra bricks sandwiched between rows, offsets, and misalignments. Defects of exactly the same kinds abound in crystals.

Defects of either kind may profoundly alter the characteristics of a crystal, sometimes for the good, sometimes not. As an example, crystals of sodium iodide, used to detect gamma rays, are deliberately contaminated with thallium to increase their light output. Without such contamination they would be almost useless.

MULTICRYSTALLINE SUBSTANCES: Many solids consist, not of a single large crystal, but of many tiny ones bound together. For example, properly made fudge contains countless sugar crystals too fine to be perceived as such by the tongue. However, if the maker was clumsy and permitted the crystals to become too big, then the fudge takes on an unpleasant grainy texture.

METALS: Metals (which for the most part are microcrystalline in structure) are among the most important solids in nature, at least technologically. They typically consist of innumerable crystals stuck tightly together. It is possible though difficult to produce single large metallic crystals. These have scientific interest but few practical applications.

When metals are rolled, hammered, tempered, or otherwise handled, all sorts of geometric defects are introduced thereby altering the metal's hardness, springiness, strength, and other properties.

As a class, metals have several unique properties; they conduct heat and electricity many times better than non-metals; they are extraordinarily ductile, malleable, and flexible; and they have a characteristic sheen. What causes these properties?

The neutral atom of any metal has one, two, or three loosely attached electrons. As these can be very easily removed, some of them drift freely within the crystal volume. It is immediately clear why a metal is a good conductor of electricity. The free electrons can move about within a wire as easily as water flows in a pipe. These free electrons are also responsible for the other properties mentioned, but the mechanism isn't quite so obvious.

ORGANIC COMPOUNDS: Carbon is unique among the elements in that it can form molecules containing dozens, hundreds, or thousands of atoms. Such molecules always contain many hydrogen atoms and may or may not contain other kinds of atoms as well, oxygen and nitrogen being the most common. In nature, such complex molecules are nearly always produced by living organisms, hence the name. Many organic compounds can exist only as solids. If enough heat is applied to melt or evaporate them, they will decompose.

Some kinds of organic molecules polymerize easily; that is, they join together like strings of freight cars. It is entirely possible for this process to go on until the entire mass is one enormous molecule. It may be startling to think of a plastic plate or a drinking cup as a single giant molecule, but it is entirely possible. Moreover, if one were to put an electrical charge on it by rubbing it with a piece of silk or fur, it would become a single huge ion.

COLLOIDS AND EMULSIONS: Colloids and emulsions are important in biology and medicine. They may be roughly defined as follows:

> A *colloid* consists of fine solid particles suspended in a liquid. An *emulsion* consists of fine droplets of one liquid dispersed in another.

Obviously the suspended particles in either case must be insoluble in the supporting liquid. To be classed as a colloid or an emulsion, the suspended particles must be so small that they settle out very slowly or not at all under normal conditions.

Colloidal particles may be either single very large (organic) molecules or groups of molecules, either organic or inorganic. Colloidal particles are usually between 10^{-7} to 10^{-5} centimeters in diameter; that is to say, ten to a thousand times bigger than an atom.

Chapter VII ATOMIC STRUCTURE

INTRODUCTION

THE TERM atomic structure is misleading because—as usually used—it refers only to the arrangement of the electrons within an atom. The composition of the nucleus is not considered. It is regarded as a point located at the center of the atom having mass and a positive electrical charge. For refined calculations one must recognize that the nucleus also has angular momentum, a magnetic moment, and other properties. However, its mass and charge alone are sufficient to explain most atomic behavior.

All that does not mean that nuclear structure is absent or uninteresting. Quite the contrary. However there are good reasons for studying it separately. For one thing, atomic and nuclear configurations—involving as they do, different types of particles and different forces—have little in common. For another, the electronic arrangement of the atom has almost nothing to do with its nuclear properties; conversely the location of the protons and the neutrons within the nucleus has scarcely any influence on atomic phenomena.

As long as atoms were objects of philosophical speculation, and even after their existence was first attested by experimental evidence, the question of their structure did not and could not arise. The reason was that if atoms were truly *atomic* in the Greek sense, they could not have any structure at all, for the very word *structure* implies the existence of component parts.

Soon however, chemists had to abandon the notion of truly

structureless atoms. Otherwise how does one account for the uniquely different properties of the various chemical elements, or how is it possible for two or more atoms to join together to form a molecule? Not only did the facts of chemistry demand an atomic structure of some sort, but even gave clues as to what that structure must be.

Let us review briefly one of the amusing early theories of atomic structure. Chemists knew that one atom of fluorine will combine with one atom of hydrogen to form hydrogen fluoride; one atom of oxygen will combine with two atoms of hydrogen to form water; one atom of nitrogen will combine with three atoms of hydrogen to form nitrogen hydride; and one atom of carbon will combine with four atoms of hydrogen to form methane. This behavior can be neatly explained by assuming that each hydrogen atom has a single projecting hook and that the atoms of fluorine, oxygen, nitrogen, and carbon have respectively one, two, three, and four eyelets on their surfaces. See Figure 6.

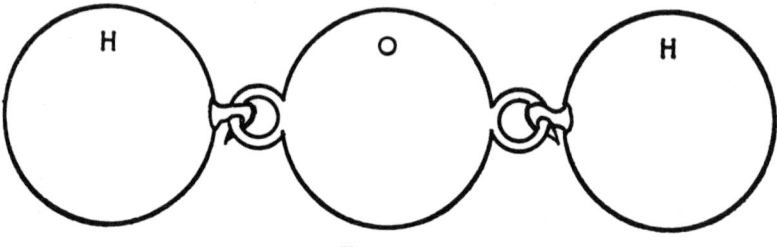

Figure 6.

Toward the end of the nineteenth century, pertinent data about atomic structure began to accumulate. At first this information was fragmentary and led to fantastic inferences now best forgotten. However as time passed, the following facts emerged:

1. The atom is more or less spherical in shape. All atoms from the lightest (hydrogen) to the heaviest (uranium) are nearly the same size with a diameter of about one angstrom (10^{-10} meter).

2. The atom contains as essential components, tiny particles called electrons which can be removed with surprising ease. (So much for philosophical indivisibility.)

3. At the center of the atom is a body called the nucleus. It has a diameter about ten thousand times smaller than the atom, contains almost all of the atomic mass, and has an electrical charge of $+Z \times 1.6 \times 10^{-19}$ coulomb.

4. The number Z mentioned above is a positive integer and is the same for all atoms of a given element. It is equal to the position which the element occupies in the periodic table, to wit: *one* for hydrogen, *two* for helium, and so on to 92 for uranium.

5. The number of electrons within an atom, represented by the letter z, is variable but approximately equal to Z. If Z and z are equal, then the negative charge is just equal to the positive charge and the atom is electrically neutral. If z is less than Z, the atom is called a *positive* ion; if greater, a *negative* ion.

6. The chemical behavior of the atom is determined both by Z and by z. Consider the element mercury for which Z is equal to eighty. There are three frequently found values of z: 78, 79, and 80. The doubly ionized ion ($z = 78$) is found in the so-called mercuric compounds, all of which have related properties. The singly charged ion ($z = 79$) is found in the mercurous compounds which are similar to one another but quite different from the mercuric compounds. Curiously, most of the mercurous compounds are violent poisons while the mercuric compounds are not. The neutral atom occurs in mercury vapor and probably in the liquid and solid phases as well.

7. The atom can exist in a number of different energy states. That state having the least possible amount of energy is called the ground state and will last indefinitely unless energy is forced into the atom from some outside source. All the other states are called excited states, and if the atom is in any one of them, it will usually return to the ground state by ejecting the surplus energy as a photon.

THE PLANETARY MODEL

These facts being established, the question arises, are the electrons arranged in any particular pattern within the atom or are they merely jumbled together? Consider the atoms of, say, oxygen. These are normally in their ground state. Since chemical and physical experiments show that they all have identical properties, it follows at once that all must have exactly the same electronic configuration.

Now let a number of these atoms be raised to their first excited state. These will now differ markedly in their properties from those atoms remaining in the ground state but will be identical to one another. We conclude that the first excited state corresponds to a slightly less stable electronic configuration. The same can be said for the second, third, and all the higher excited states.

Since the atomic electrons must be arranged in patterns, what are those patterns? There were several proposals, but the one first showing signs of success was the planetary theory. According to it, the electrons revolve around the nucleus just as the planets revolve around the sun. That sounds plausible enough since the operative forces, gravitational and electrostatic, have the same mathematical form. The great flaw with the planetary theory is that it permits the system to have any quantity whatever of energy, and even worse, there is an unlimited number of possible planetary arrangements for each energy value.

This obstacle was by-passed by arbitrarily assuming that only a discrete set of orbits is permissible, and that only a limited number of electrons can occupy any particular orbit. At that time no reason could be given for such restrictions except that they were necessary to fit the experimental facts.

Figure 7 shows the arrangement that works. The black dot in the center represents the nucleus, and the successive circles are the electronic orbits. There is theoretically an infinite number of these, but only the first dozen or so are of any consequence. These orbits are conventionally designated in one of two ways: they may be numbered 1, 2, 3, and so on from the inside out or they may be labeled K, L, M, N, and so on, again beginning with the innermost orbit. The latter system is used in the figure. The numbers in the figure show the maximum number of electrons that orbit may contain.

For the first eighteen elements, namely those from hydrogen through argon, the ground state is formed by placing all the

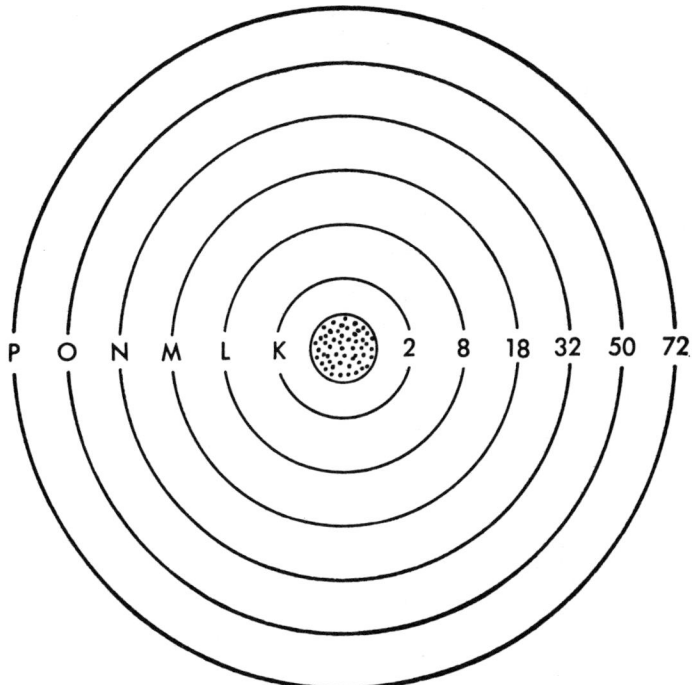

Figure 7.

electrons as near the nucleus as possible. The first two go into the K orbit, the next eight into the L orbit and so on. Beyond argon this rule breaks down for reasons that do not concern us at the moment.

The electronic orbits are often called shells. The reason is that the electrons in any particular orbit are not confined to a plane but may move anywhere over the surface of a sphere centered on the nucleus and having the same radius as the orbit. Returning to the solar system which was the original inspiration for the planetary model of the nucleus, we observe that the planets do in fact move in orbits all of which very nearly lie in the same plane so that the system as a whole resembles a pancake. This fact, however, is not a necessary consequence of physical law, but is due to the way in which the system was created. As these conditions are not relevant to the atom, the electrons can and do move through three-dimensional space.

We now know that the planetary model of the atom is not a photographic representation as some originally thought. In fact no picture of any sort is possible because the atomic electrons have no assignable position or velocity. Any electron may be found anywhere within the atom or even a short distance outside; it may be moving in any direction at any reasonably low speed.

This statement seems to contradict what was said earlier, that the contents of the atom cannot be randomly jumbled but must have definite arrangement. Our previous statement is correct; it is our notion of structure that is wrong. When speaking of the atom, structure does not mean position and velocity but rather a set of well defined states each characterized by four physical quantities; *energy, angular momentum, magnetic orientation,* and *spin direction.* Of these only energy is immediately important for our purposes.

Therefore when we say that a certain electron is in the K shell, we do not mean that it is actually moving in such and such a circle about the nucleus but that it has a particular set of values for the four quantities above mentioned. In particular, the K electrons have the lowest possible energy, the L electron the next lowest, and so on.

The planetary model of the atom does have some pictorial value however. While the K electrons may be found anywhere, they are most likely to be in a region near the nucleus. The L electrons will usually be found a little further out, and the M electrons still further out and so on.

Although no realistic picture of the atom is possible, the orbital model is still extremely useful, if one remembers that it is symbolic, not literal. It may be compared to the circuit diagram of a television set. While such a diagram bears no visual likeness to the actual circuitry, each symbol in the drawing represents a particular component of the circuit and its interconnections with the others.

CHEMICAL PROPERTIES

As mentioned earlier, many chemical properties of the elements can be explained by assuming that the atoms have one or more hooks or eyes attached to them so that they can link up with other atoms. Probably no one took this notion very seriously, but it was too convenient and useful to ignore. In the light of present knowledge, the question arises, how can a structure that resembles the solar system link up with other, similar structures at all? Furthermore, how is it that some atoms can combine with only one, others with two, and still others with three or more? A complete answer to this question would fill several highly mathematical books, but a simple explanation is possible for many phenomena.

As far as chemical behavior is concerned, only the electrons in the outermost nonempty shell are involved. The binding energies of the other electrons are much too high to be disturbed by chemical means. The outer electrons, called *valence electrons*, have binding energies of only a few electron volts, the exact figure varying from element to element.

In a few elements the valence shell is filled. These are the noble gases, *helium, neon, argon, krypton, xenon,* and *radon*. For several reasons a filled shell is an especially stable configuration. Consequently atoms so equipped have very little desire to interact with other atoms. It was once thought—and a number of books still state—that the noble gases can not combine chemically with anything. It is now known that the heavier ones may form a few compounds under certain circumstances.

Another class of elements consists of those which in the neutral state have just one valence electron. As this electron is at some distance from the nucleus and is effectively screened from the latter's positive charge by the closer lying electrons, its binding energy is very low. These elements are called *alkali metals* and include *lithium, sodium, potassium,* and *cesium*. The valence electron being lightly bound, alkali atoms are very likely to lose that electron and hence exist as positively charged ions. Accordingly the sodium atoms in common salt (as well as in other sodium compounds) are all ionized.

The *halogens* constitute a family of elements complementary to the alkalis. In these elements the valence shell is just one short of being filled. The chief members are *fluorine, chlorine, bromine,* and *iodine*. These atoms have an especial affinity for electrons, so that they are usually found as negative ions, having captured an electron from the environment.

Since the electronic configuration of both the negative halogen ion and the positive alkali ion is exactly the same as that of the neighboring noble gas, one might suppose that they would be chemically neutral. That is not true however for the ion as a whole has an electrical charge. Such ions therefore strongly attract others of the opposite sign.

The alkali metals and the halogens are especially likely to interact with each other. Whenever two such atoms come near one another, the loosely bound electron of the one is caught into the vacancy of the other thus creating a positive-negative pair. See Figure 8. Thereafter several things may happen. If the two

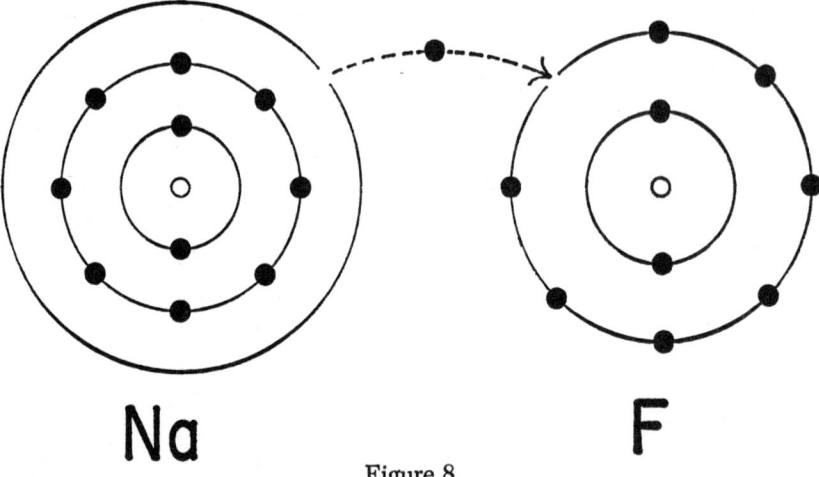

Figure 8.

ions are isolated in empty space, they will stick together as a diatomic molecule. If a number of such atoms are permitted to combine near one another at low temperature, they will organize themselves into a crystal, of which common table salt (NaCl) is an excellent example. If these ions come into being in water, an ionic solution will result. Salt as such does not exist in solution. Instead, there are two populations of ions each moving independently of one another. However, because of their opposite charges, they remain thoroughly intermixed. Otherwise extremely strong electrostatic forces would arise. When the solvent is evaporated, the ions reorganize themselves into the crystalline configuration of sodium chloride.

Atomic Structure

The phenomenon we have been describing is called *ionic binding*, a process whereby very stable chemical compounds are made. The necessary ingredients for such compounds are two kinds of atoms, the first with one or two weakly attached electrons that can be easily removed, and the other with one or two vacancies in the outer shell and therefore avid for extra electrons. From what we have seen it is clear that the alkali and the halogen atoms combine just as if the atoms of one had a hook each and the atoms of the other an eyelet.

A more common but usually less stable type of interatomic connection is the *covalent bond*. In such bonds neither atom is ionized; each shares one or more electrons. The water molecule is a common and typical example.

Oxygen has two vacancies in the outer shell and can therefore accept two additional electrons. However this would give the resulting ion two extra units of negative charge which would tend to disrupt it. Hence oxygen is considerably less likely to form ionic bonds by electronic capture than are the halogens.

Hydrogen is unique among the elements in that it has both one electron and one vacancy in the valence (K) shell. Consequently it may act curiously like either an alkali or a halogen or like neither. What happens when hydrogen and oxygen combine to form water is illustrated in Figure 9. The hydrogen atom

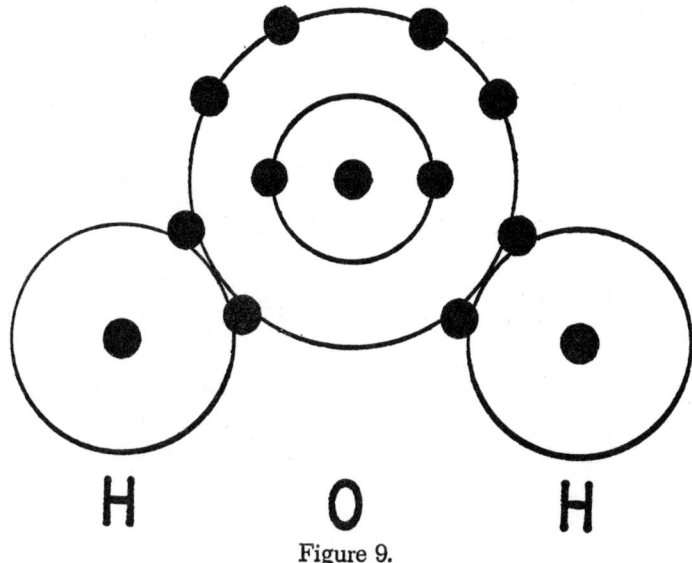

Figure 9.

moves near the oxygen atom so that the one electron of the former more or less effectively fills one vacancy in the latter without however completely breaking its initial tie. Likewise an adjacent electron of the oxygen atom also fills the vacancy in the hydrogen atom without becoming totally detached. Thus by each sharing an electron with the other, the vacancies of both are filled. Since oxygen has two vacancies, it can simultaneously share electrons with a second hydrogen atom thus giving the familiar molecular formula H_2O.

The covalent bond therefore accomplishes the same thing as the ionic bond, namely that of giving each atom involved a closed-shell structure. However the means are quite different. Re ionic bonding, the one or two outer electrons of one atom are stripped off to fill the corresponding vacancies in the other; there is a permanent displacement of the electrons in question from one atom to the other. In covalent bonding on the other hand, the chemically active electrons belong to neither atom but are simultaneously joined to both.

We now see why different atoms are capable of making different numbers of bonds. Since oxygen has two vacancies, it can form two bonds, either with two different atoms or with a single atom also capable of making two bonds. Calcium, which has two valence electrons, can also form two bonds either by sharing or by donating those two electrons to another atom. The extraordinary chemical activity of carbon is in part due to the fact that it has four electrons in a shell capable of containing eight. Hence it can form four bonds with as many as four different atoms.

THE NEUTRAL HYDROGEN ATOM

The neutral hydrogen atom (which can exist only in free space) has a single positive charge on its nucleus and a single orbital electron; that is, $Z = z = 1$. If the atom is in its ground state, the electron is in the K shell. While the position of this electron is uncertain, its energy is precisely -13.6 eV. This is the minimum amount of energy required to remove the electron from the atom.

The electron could, of course, be in any other shell, each of which also has a definite energy associated with it:

K shell .... −13.6 eV
L shell .... −3.39 eV
M shell .... −1.50 eV
N shell .... −0.85 eV and so on

There is an infinite number of shells, each with an energy a little nearer zero than the preceding. In every case the energy associated with the shell is just that required to remove the electron completely.

Suppose that one wants to move the electron from the K to the L shell, that is, to the first excited state. The energy required is the energy of the initial state minus the energy of the final state:

$$-13.6 - (-3.39) = -10.21 \text{ eV}$$

The meaning of the minus sign is that energy must be fed into the system as mentioned earlier.

The electron will not remain in the L shell but will soon return to the K shell and in so doing will give up the energy required for the original transfer. This energy will usually come off as a photon having—in this case—10.21 eV. Photons of this energy are in the ultraviolet region.

There are several ways to excite an atom, the most common being bombardment with photons or electrons. If a photon having exactly the right amount of energy to effect the transfer comes near the electron, it will probably knock that electron into the higher orbit and vanish in the process. If electrons are used, those electrons must have at least as much as, and preferably somewhat more than, the energy required for the excitation.

Suppose that the atom had been raised, not to the first excited state, but to a higher one, say, the third. It may now return to the ground state by any of several routes. It may go directly, and in so doing, radiate, a photon having an energy equal to the difference of the energies of the third excited state and the ground state:

$$\text{Energy of Photon} = -0.85 - (-13.6) = +12.75 \text{ eV}$$

The electron may also return one level at a time or it may return by skipping some intervening levels and stopping at others. The number of photons given off is equal to the number of stages into which the return journey is divided. The energy of each photon is equal to the difference of the energies of the levels concerned.

ENERGY LEVEL DIAGRAMS

Since energy is far more important than position for atomic electrons, an energy level diagram is more useful than a sketch of the electronic shells. A simplified diagram for hydrogen is shown in Figure 10.

Each short horizontal line represents the energy of a particular shell; the spacing between successive lines is proportional to the energy difference. The distance between any line and the line AB is proportional to the energy required to remove an electron in that shell completely from the atom. The letters to the left indicate the orbit to which the line belongs; the number to the right is the energy in electron volts of that level.

The two vertical arrows represent a two-stage transition of an electron from the N shell via the M shell to the ground state. The length of each arrow is a measure of the energy released in the jump. If that energy is given off as a photon (as is probable), then the length of the arrow is also a measure of the photon's energy.

Hydrogen atoms seldom occur unattached in nature. Either they are combined with atoms of other elements or are joined up in pairs to form molecules of hydrogen gas. (Elemental hydrogen can also exist as a solid or liquid but only at extremely low temperatures). Even if separate atoms were originally present, they would pair up to form diatomic molecules as soon as they were brought together.

There are two reasons for such behavior. First, the atom "dislikes" a partly filled shell. If there are only one or two electrons in the outer shell, the atom is likely to discard them whenever it can reasonably do so; if the outer shell lacks one or two electrons of being full, the atom will seize enough from the environment to fill it out. Not only do atoms dislike partly filled

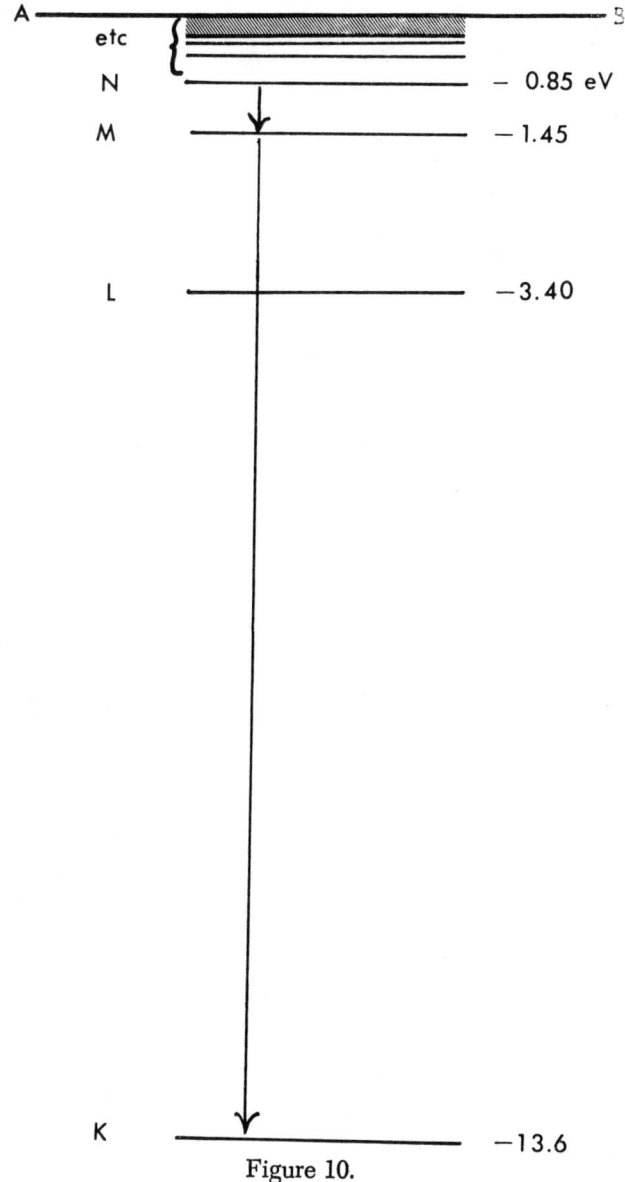

Figure 10.

shells; they "abhor" unpaired electrons. Hence, if an atom finds itself with an odd number of electrons, it will either dispose of one or acquire another to even up the count.

Clearly the neutral hydrogen atom is in a most unhappy state;

its K shell is only half full (or half empty) and the total number of electrons is odd. When two hydrogen atoms come near one another, all would be set right if one atom should rob the other. The resulting pair of ions, one positive and the other negative, would then stick together because of the electrostatic attraction between them to form a stable diatomic molecule.

However this process is impossible. Since the atoms are identical, there is no way to decide who is to be the robber and who the victim. Since the K shell is either half full or half empty, each atom is just as ready to give up an electron as to take on another one. There is a way out though. If the two atoms move together so that each electron is simultaneously in the K shell of both atoms, thus forming a covalent bond, then both are relatively "happy."

THE NEUTRAL HELIUM ATOM

The neutral helium atom has two positive charges in its nucleus and two orbital electrons which just fill the K shell. Since the electronic shell is closed and the number of electrons is even, the helium atom is in the best of all possible states. Therefore, it does not associate chemically with other atoms, either of helium or of any other element. There are no helium compounds, and elemental helium is an agglomeration of individual atoms.

THE SECOND-ROW ELEMENTS

The next row of the periodic table contains eight elements, those whose outermost electrons are found in the L shell. They are:

Element	Z	Number of electrons in the L shell	Number of vacancies in the L shell
Lithium	3	1	7
Beryllium	4	2	6
Boron	5	3	5
Carbon	6	4	4
Nitrogen	7	5	3
Oxygen	8	6	2
Fluorine	9	7	1
Neon	10	8	0

Atomic Structure

The number of electrons and vacancies in the above table apply to the neutral atom, not to any of its possible ions.

THE LITHIUM ATOM: If the helium atom is supremely happy, the lithium atom is utterly miserable; it has an unpaired electron in an otherwise empty orbit. As it will dispose of this odd electron if at all possible, there are very few neutral lithium atoms about; instead, most of the lithium on earth is in the form of positive ions.

THE FLUORINE ATOM: The fluorine atom lacks just one of having a full shell of electrons and, hence, is just as miserable as the lithium atom. It will therefore seize an electron on the first opportunity to fill out the L shell and will thereafter exist as a negative ion. In particular, if a fluorine atom and a lithium atom come near one another, the first will appropriate the electron that the other is all too willing to give up. The two ions will henceforth remain together as a molecule of lithium fluoride.

THE CARBON ATOM: Located exactly in the middle of the row is carbon with four electrons and four vacancies. Since the number of electrons is even and the atom is undecided as to whether it should try to pick up additional electrons or let some go, it is reasonably content to remain in the elemental state. Accordingly there are large deposits of it as coal and graphite on earth, as well as much smaller ones of diamond. Carbon does however form compounds easily by giving, receiving, or sharing electrons as appropriate. Because of this flexibility and because the binding forces of all atoms with so few electrons tend to be strong, carbon can form an amazing number of compounds with molecules containing hundreds and thousands of atoms.

THE OTHER SECOND-ROW ELEMENTS

Beryllium has two electrons to give or to share and boron three. Of these two beryllium is metallic while boron is transitional.

Oxygen and nitrogen have respectively two and three vacancies and can form ionic or covalent bonds accordingly.

Neon with its filled L shell is a noble gas.

SHELLS AND SUBSHELLS

It is not quite true that the energy of all the electrons in any given shell is the same. All of the shells except the K are divided into subshells, each with an energy of its own. Usually the differences existing between subshells are considerably less than those between the main shells.

As one proceeds from the nucleus, each shell has one more subshell than the one preceding. The K shell is undivided; the L shell has two subshells; the M, three; the N, four; and so on. For historical reasons the successive subshells within each main shell are labeled:

$$s, p, d, f, g, h, i, k, l, \ldots$$

From k on, the subshells are labeled alphabetically except that the letters p and s must be skipped, having already been used. The letter j is omitted since the subshell labeling was first done by German physicists. In their language the letters i and j are not considered distinct.

The K shell is simultaneously an s subshell; the L shell contains an s and a p subshell; the M shell contains an s, p, and d subshell and so on. See Figure 11. The maximum number of electrons which each subshell may contain varies:

   an s subshell may contain 2 electrons
   a p subshell may contain 6 electrons
   a d subshell may contain 10 electrons and in general,
   the nth subshell may contain $2(2n-1)$ electrons

Within a main shell the s electrons have the highest (most negative) binding energy, the p electrons a little less, and so on out. Beyond argon in the periodic table, the energy values of the outermost subshells of one main shell overlap the s and perhaps the p subshells of the next. This fact explains the chemical irregularities found among the heavier elements.

METALS AND INSULATORS

When elements having but few valence electrons are in the solid state, some of these electrons do not remain bound to their respective atoms but wander throughout the volume of the

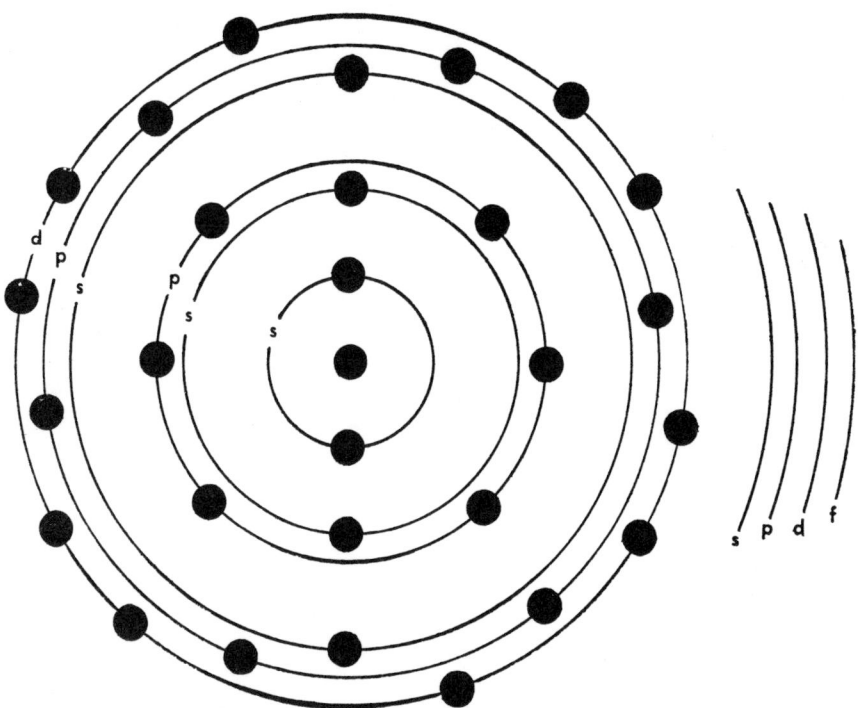

Figure 11.

substance. These *free electrons,* as they are called, are responsible for the three distinctive properties of metals: a mirror like sheen, electrical conductivity, and high heat conductivity.

Elements with the outer orbit over half full are nonmetallic. Those in the first two rows of the periodic table are gases except at very low temperatures; further down they are usually solids at ordinary temperatures. In these materials the electrons are all firmly bound to the parent atom since such atoms are hungry for more electrons than they possess and are not disposed to let any go. Consequently the properties of these substances are the reverse of metals. They do not have a metallic luster, they are good electrical insulators, and they transfer heat but poorly.

Those elements lying between the metals and the nonmetals in the periodic table—that is to say, those whose outer orbits are about half filled—are transitional in character. Carbon, for example, although being quite nonmetallic in texture and appearance, is nevertheless a fair conductor of electricity. One form of elemental carbon, graphite, shows faint hints of metallic sheen in certain lights; another form, diamond, is totally non-metallic in its properties. These in-between elements are usually called semiconductors or sometimes metalloids. Formerly, they were of little importance (apart from carbon) but now their unique properties have made them invaluable in the electronics industry.

Chapter VIII NUCLEAR STRUCTURE

NUCLEAR MODELS

IT HAS NOW BEEN established beyond reasonable doubt that the atomic nucleus is an assembly of neutrons and protons. Other particles, called *mesons*, are believed to be continuously created and absorbed within the nuclear volume, but their existence is not necessary to our discussion. Henceforth we shall ignore them.

The nucleonic arrangement is very different from that of the atomic electrons. The latter are spaced far apart, the distance between them being over a thousand times greater than their diameters; the nucleons however are closely packed. If one compares the volume of any nucleus with that of the nucleons within it, one sees at once that the nucleons are almost touching. Since these particles are moving at high speed, they must repeatedly collide with one another so that orderly arrangement is impossible. This fact alone makes the study of the nucleus far more complex than that of the electronic configurations.

The forces between nucleons in a nucleus and the molecules of a liquid are roughly comparable. In the latter, the intermolecular force behaves somewhat as follows: If the distance between two molecules is more than about one atomic diameter (10^{-8} cm), the force is ineffective. As the molecules approach one another, the force becomes strongly attractive so as to draw them still closer. Once they are—so to speak—touching, any attempt to shove them still closer will be countered by a strong repulsive force such as one experiences when trying to squeeze two tennis balls together tightly enough to deform them. Accordingly, there is a preferred distance from center to center for molecules in a liquid. If they are a little too far apart, they will be drawn nearer; should they come a little too close, they will be driven apart.

The nuclear force behaves in much the same way. It is practically zero as long as the distance between nucleons is more than about one nucleonic diameter (10^{-13} cm). Then it becomes exceedingly strong until the particles are *in contact*. Attempts to compress or deform them will be firmly resisted.

Clearly the nucleus must resemble a drop of liquid, the principal difference being that the number of particles in the nucleus varies from one to about three hundred while the number of molecules in a rain drop is in the sextillions. Nevertheless physicists have had considerable success in explaining the behavior of the larger nuclei by treating them as drops of liquid; so much so that one speaks of the *liquid drop model* of the nucleus.

A question arises: Since—as the oceans prove—a liquid may attain enormous volume, why do not nucleons also accumulate in huge masses?

The answer is that a second force, disruptive in effect, prevents large clusters of nucleons from occurring. This force is the electrostatic repulsion between protons. In a liquid, the positive charges of the atomic nuclei are exactly balanced by the negative charges of the surrounding electrons so that no net charge exists. Even in those liquids containing ions, the positive just counterbalance the negative so that the fluid as a whole is electrically neutral.

In the nucleus however, there are no negative charges to offset the protons. Furthermore, the electrostatic force has a long range; every proton in a nucleus is repelled by every other while it is attracted by only the few nucleons in its immediate vicinity. Hence in the larger nuclei the disruptive effect becomes proportionately greater. The result is that no nucleus with more than 208 nucleons can be permanently stable. In larger ones the electrostatic repulsion will eventually succeed in breaking some of the bonds and thus bring about a nuclear disintegration.

Most heavy nuclei lose electrostatic charge by alpha emission. In this process two protons and two neutrons join together as a unit and leave the nucleus which is thereupon lighter by four nucleons than it was before. If the loss of an alpha particle brings the total nucleon count down to 208 or less, the nucleus may thereafter be stable; otherwise another alpha particle must be ejected sooner or later. Some nuclei with A equal to or greater than 232 may slim down by splitting into two approximately equal fragments. This latter process is called *nuclear fission*.

Although the liquid drop model explains many nuclear phenomena, it hopelessly fails for others. These failures cannot be attributed to the small number of particles present or to the disruptive electrostatic force.

Amazing as it may seem in view of what has already been said, many well established facts clearly imply that the nucleons must after all move in orbits much like the extranuclear electrons. To account for these facts, a whole theory based on the *Nuclear Shell Model* has been developed. Like the liquid drop model it explains certain things but not others.

The truth is doubtless somewhere in between. The nucleons try to pull themselves into a regular orbital pattern which, however, is continually broken up by the many collisions. This state of part order, part disorder accounts for the seemingly contradictory behavior of nuclei.

NUCLEAR ENERGY LEVELS

The nucleus has excited energy states just as do the atomic electrons, but because of the complex behavior of the nucleons, the positions of the nuclear states are much more irregular and unpredictable. Whereas the energies of the electronic levels can often be calculated with great accuracy, those of the nuclear levels must be determined by experiment.

An excited nucleus will usually return to the ground state by emitting a photon. In most cases, the transition occurs within about 10^{-13} sec, but there are a few excited states that may persist for minutes, hours, or days. These latter are called *metastable* or *isomeric* states and are of great importance both theoretically and practically. A nuclide with all of its nuclei in such a state is called an isomere of the nuclide in question.

The existence of excited nuclear states indicates that the nucleons must occupy definite energy levels just as the electrons do. Where the latter are concerned, the energy states can be associated with definite orbits; in the nucleus such identification is only partially possible as we have seen. However two facts concerning nucleon *orbits* are well established:

1. Each may contain zero, one, or two protons and zero, one, or two neutrons. A proton can not fill a neutron vacancy or vice versa.
2. Each orbit has a definite binding energy which is equal to the work required to remove a nucleon from that orbit to a position completely outside the atom. There may be and usually is some difference in the energy required according as the particle in question is a neutron or proton.

It is not worth while trying to sketch nuclear orbits as one does for the electrons since the result would have no pictorial validity. A diagram as shown in Figure 12 can be very useful however. The horizontal lines represent the orbits; the distances between them and the long dotted line are proportional to their respective binding energies. The vertical line divides each orbit into halves; one for the protons and one for the neutrons. The black dots indicate nucleons that are present. In the figure, the lowest orbit is filled; the second has its full quota of neutrons but only one proton. The higher orbits are all empty. This particular figure applies to the Li-7 nucleus in its ground state.

The preceding diagram is suitable only for very light nuclei. If a nucleus contains several protons, their mutual electrostatic repulsion makes them somewhat less tightly bound than the corresponding neutrons. This fact can be represented by placing a jog in the center of each line as shown in Figure 13.

If all the nucleons are in the lowest available orbits, the nu-

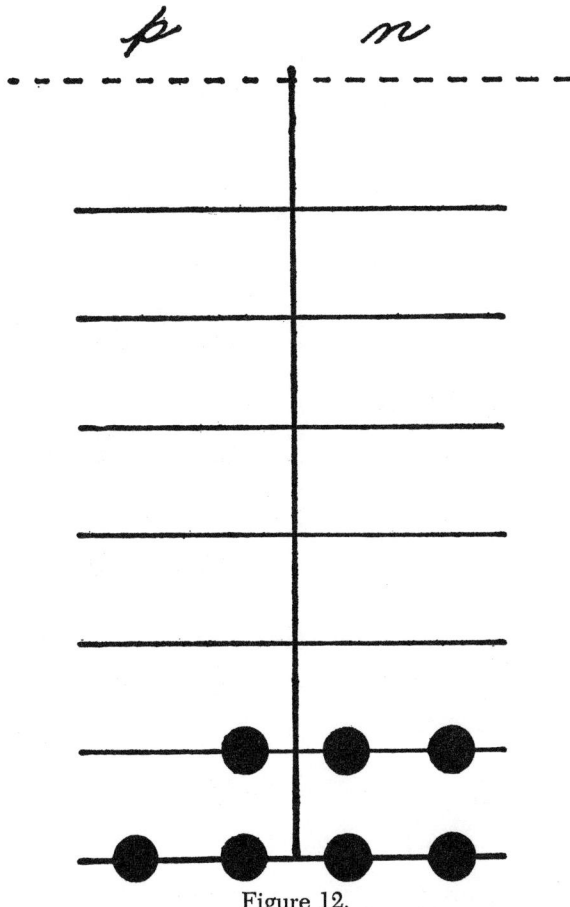

Figure 12.

cleus is in its ground state. If one or more is in an unnecessarily high orbit, the nucleus is excited. A nucleus will return to the ground state in much the same way as an atom. The nucleon will return to the lowest available orbit in one or several leaps giving up energy each step of the way, usually in the form of a photon. This process is shown in Figure 14.

There are some significant differences between nuclear and atomic de-excitation:

1. Whereas the energy differences in atomic transitions would never exceed 150 keV and are often much less, the energy differences in nuclear transitions usually lie between 100 keV and 5 MeV. Values outside these limits are possible.

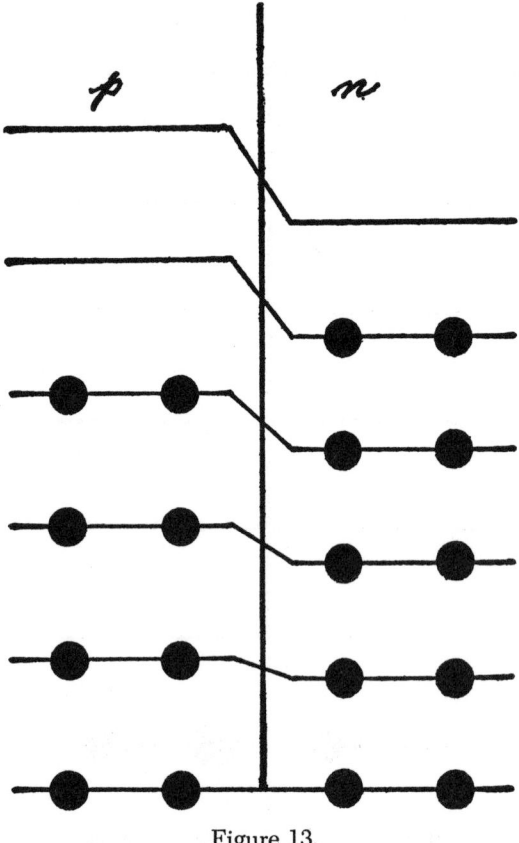

Figure 13.

2. Nuclear transitions usually occur about 10,000 times faster than atomic transitions.
3. Metastable states are much more important in nuclei than in atoms. Some isomeric nuclides such as technecium-99m are of great value in medical practice and research.

NUCLEAR LEVEL DIAGRAMS

In most cases we do not have enough information to make orbital diagrams for the excited states of the nucleus. Since nuclear properties cannot be calculated in most cases, our only knowledge about the excited states in nuclei is that which can be obtained experimentally. To take a particularly simple case: Carbon-12 can be excited to give off a gamma ray of 4.44 MeV.

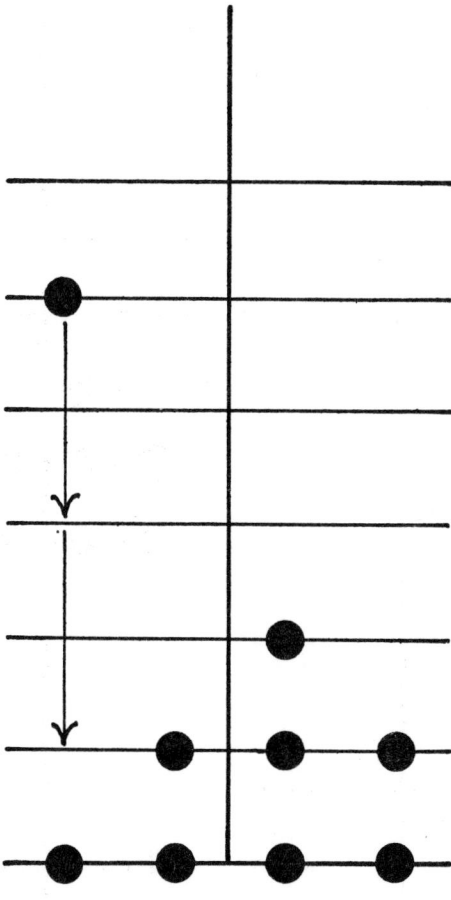

Figure 14.

This fact implies the existence of an excited state of 4.44 MeV above the ground state. Further experiments indicate that this is also the lowest excited state. These facts can be expressed in the diagram shown in Figure 15a. The arrow indicates that the excited nucleus returns to the ground state by gamma ray emission.

Experiment does not tell us whether that state corresponds to a raising of one proton, one neutron or perhaps more than one nucleon to a higher level. The correspondence between the experimental diagram and the possible orbital configurations are shown in Figure 15b–d.

72 Basic Nuclear Physics for Medical Personnel

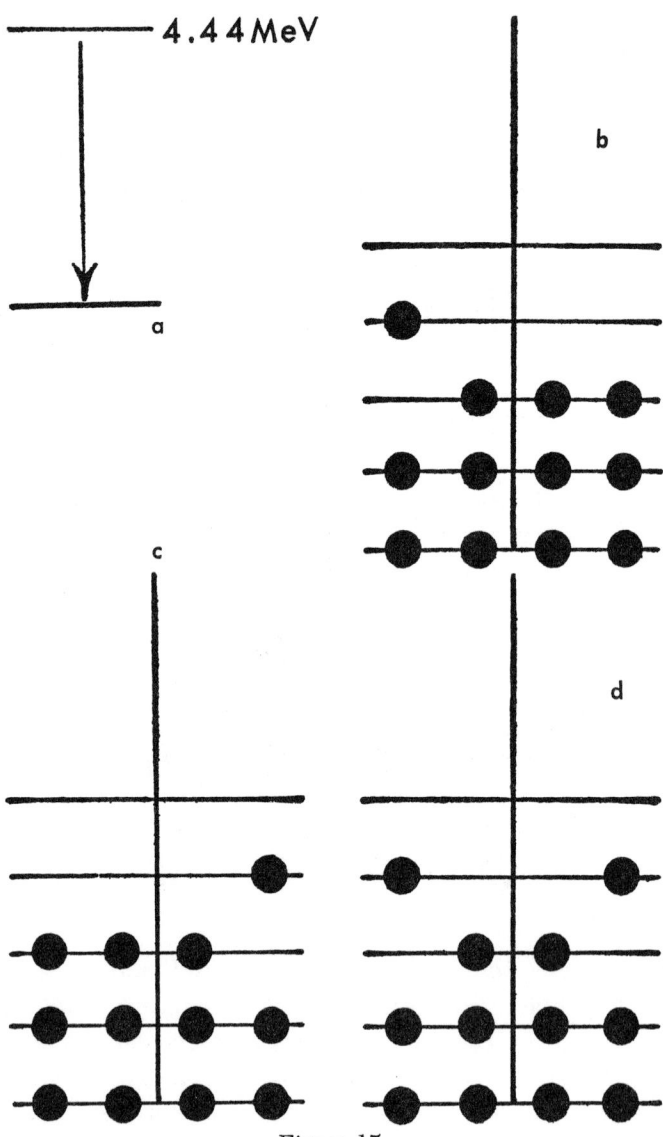

Figure 15.

Figure 16 is an energy level diagram for technecium-99. The heavy horizontal line at the bottom represents the ground state while the higher lines represent excited states that have thus far been identified. Contrary to the convention applied to atomic levels wherewith all bound electronic configurations have nega-

Nuclear Structure

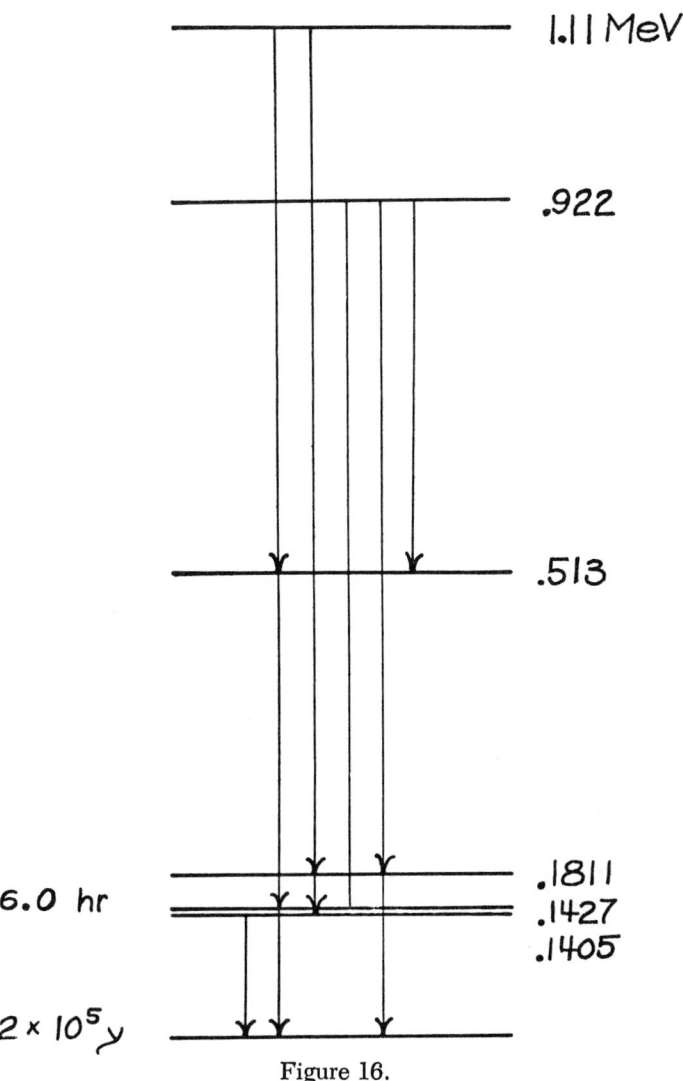

Figure 16.

tive energies, the ground state of the nucleus is assigned the value zero. Accordingly the excited states must have positive energy values. These are listed just to the right of each line. The notation 6.04 hr to the left of the second level means that it is metastable with a half life equal to the value given. The other levels, as far as is known, decay at the usual rate.

The vertical arrows indicate the known modes of decay from

the various excited states. Note that each will decay to some but usually not to all of the states below it. The rules governing these preferences are known but are beyond the scope of this book.

The energy lost in a given transition is equal to the difference of the binding energies of the initial and final states. The energy is usually carried off by a gamma ray, but sometimes by a conversion electron.

PROBLEMS

1. Assume that each of the transitions shown in Figure 16 leads to the emission of a gamma ray. What is the energy of each?

2. Draw orbital diagrams similar to the one shown in Figure 12 for H-2, He-4, C-13, N-14, O-16, and F-19. Which of these have unpaired nucleons?

3. The radius of a nucleon is about 1.4×10^{-13} cm. The radius of the Al-27 nucleus is 4.71×10^{-13} cm. What per cent of the nuclear volume is occupied by the nucleons?

Chapter IX RADIOACTIVE DECAY

THE DIFFERENTIAL DECAY LAW

IF WE EXAMINE the nuclei of any radionuclide, all of them will react the same way to every test. They can not be identical, however, for they will decay at different times; some almost at once, some soon, some much later. There is no advance evidence that any particular nucleus is about to disintegrate. One that will be gone in ten seconds is indistinguishable from one that will last ten years.

While it is impossible to predict the fate of any one nucleus, we can—if given a large number of them—say quite accurately what fraction will decay in any specified period of time. We are in the same position as the life insurance people who know how many men between, say, 55 and 65, will die of heart attacks during the year but have no idea who the victims will be.

There are a number of different kinds of nuclear disintegration, alpha, beta, and fission being the most common. All obey the same decay law. Suppose we begin with a number N of nuclei of a certain radionuclide and observe them for a time interval Δt so short that the number decaying ΔN is much less than the total. Then we shall find that ΔN is proportional to both the number originally present and the length of the time interval:

$$\Delta N \propto N$$
$$\Delta N \propto \Delta t$$

These two proportionalities yield the equation:

$$\Delta N = -\lambda N \Delta t$$

where λ is the decay constant. λ has a definite value for each radioactive species which (with very rare exceptions) can not be changed by altering the temperature, pressure, or chemical environment to the extent possible in earthly laboratories.

Since N is decreased by radioactive decay, ΔN must be negative. That is why the minus sign appears in the equation.

This equation can not be used if the time Δt is long enough for the number decaying to be an appreciable fraction of those originally present. In that case N has no definite value that can be used for computation.

EXAMPLE: The decay constant of a certain radionuclide is 10^{-6} sec^{-1}. If a sample of this nuclide originally contains one billion atoms, how many will be left three minutes later?

SOLUTION:

$$\Delta N = -\lambda N \Delta t$$
$$= -10^{-6} \times 10^{9} \times 180$$
$$= -180,000$$

Number remaining $= N + \Delta N$
$$= 1,000,000,000 - 180,000$$
$$= 999,820,000$$

ACTIVITY

The word *activity* means the disintegration rate of a radioactive sample. Although one can specify activity by stating the number of atoms decaying per second, such is not convenient because the numbers are usually enormous. Hence a special unit called the *curie* (abbreviation Ci) has been established. It is by definition equal to 3.7×10^{10} disintegrations per second. This value was chosen because it was thought to be the activity of one gram of radium. We now know that the correct value for radium is a little less, but the definition remains.

EXAMPLE: What is the initial activity of the sample mentioned in the example above?

Solution:

$$-\frac{\Delta N}{\Delta t} = \lambda N = 10^{-6} \times 10^9 = 10^3 \frac{\text{dis}}{\text{sec}}$$

$$\frac{1000 \frac{\text{dis}}{\text{sec}}}{3.7 \times 10^{10} \frac{\text{dis}}{\text{sec-Ci}}} = 2.7 \times 10^{-8} \text{ Ci}$$

Batches of radioactive materials in common use vary greatly in size. A nuclear reactor contains millions of curies; a cobalt teletherapy unit, several thousand. Many radiation sources used in industry contain about a curie more or less. For therapeutic purposes patients will receive several thousandths of a curie. For diagnostic tests a few millionths usually suffice. In the realm of health physics, quantities as small as a millionth part of a million may be of concern. To span this range of magnitudes, the standard metric prefixes are used:

megacurie (MCi)	1,000,000 Ci
kilocurie (kCi)	1,000 Ci
millicurie (mCi)	10^{-3} Ci
microcurie (μCi)	10^{-6} Ci
nanocurie (nCi)	10^{-9} Ci
picocurie (pCi)	10^{-12} Ci

The above decay formula gives the number of atoms decaying as a function of the total number in the sample. However it can be more conveniently written with activity A as the variable:

$$\Delta A = -\lambda A \Delta t$$

EXAMPLE: The decay constant of I-131 is 0.0859 da^{-1}. If the activity of a certain sample is 5.0 mCi at noon, what is its activity at six P.M. of the same day?

$$\Delta A = -0.0859 \times 5 \times \frac{1}{4}$$
$$= -0.107 \text{ mCi}$$
$$A + \Delta A = 5.0 - 0.107 = 4.893 \text{ mCi}$$

THE INTEGRAL DECAY LAW

The previously given formula cannot be used if the time interval is long enough for an appreciable quantity of the original material to decay. For that case the more complicated integral decay formula must be used:

$$A(t) = A(0)e^{-\lambda t}$$

where $A(0)$ is the original activity, $A(t)$ is the activity at a later time t, and $e^{-\lambda t}$ is the sum of the series:

$$e^{-\lambda t} = 1 - \frac{\lambda t}{1} + \frac{\lambda^2 t^2}{1 \cdot 2} - \frac{\lambda^3 t^3}{1 \cdot 2 \cdot 3} + \frac{\lambda^4 t^4}{1 \cdot 2 \cdot 3 \cdot 4} - \cdots$$

and so on without end.

The existence of two different laws does not mean that nature is inconsistent. It is easy to show that the differential law is an approximation of the integral law just quoted. For short time intervals both give very nearly the same answer. For larger time intervals, the inadequacy of the differential law becomes apparent.

Obviously this formula cannot be readily solved by pencil and paper methods. One can however use:

a) a slide rule
b) mathematical tables
c) a computer, or
d) semilogarithmic graph paper

The last, although the least accurate, is the easiest to use and is adequate for most purposes. To prepare a graph for the function $e^{-\lambda t}$ you need to know three facts:

a) $e^0 = 1$
b) $e^{-1} = 0.369...$
c) the semilogarithmic plot of the function $e^{-\lambda t}$ is a straight line.

On a sheet of semilogarithmic paper label the linear axis t and the other $e^{-\lambda t}$. Plot the points (0,1) and (1/λ,0.369). Draw a straight line through these. To use this graph, find the appropriate value of t on the horizontal axis and read up to the corresponding value for $e^{-\lambda t}$. See Figure 17.

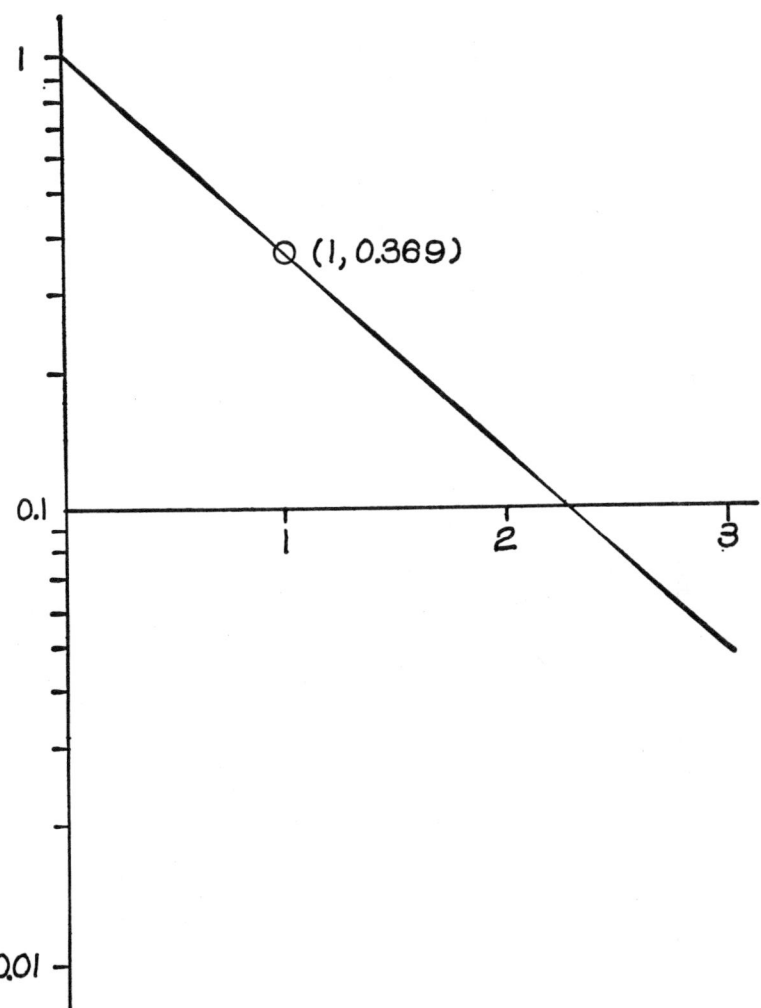

Figure 17.

HALF LIFE

With regard to radioactive materials, the term *half life* means the length of time required for one half of the atoms originally present to decay. An exceedingly wide range of half lives occurs in nature from fractions of a second to billions of years. The half lives of a few of the most important nuclides (some or all of which should be memorized according to the needs of the reader) are:

H-3	12.3. yr
C-14	5730 yr
K-40	1.3×10^9 yr
Fe-59	45 da
Co-57	270 da
Co-60	5.26 yr
Sr-87m	2.9 hr
Sr-90	27.7 yr
Tc-99m	6.04 hr
In-113m	1.7 hr
I-125	60 da
I-131	8.05 da
Au-198	2.70 da
Ra-226	1600 yr
U-238	4.5×10^9 yr

A sometimes-met misconception regarding nuclear decay is: If half of the activity disappears during the first half life, then the remainder will disappear in the second. According to this—erroneous!—logic, one curie of I-131 would be reduced to one-half curie in eight days and would be completely gone in sixteen. This conclusion is *wrong*. During the second half life only one half of the remainder vanishes so that one fourth of the original amount is left. During the third half life, one half of the remaining fourth decays so that one eighth of the original quantity is left and so on. This argument implies that any sample of radioactive material never completely disappears as the series of residues

1, 1/2, 1/4, 1/8, 1/16, 1/32, ...

never reaches zero however far extended.

That wide spread conclusion is not quite correct either, for any quantity of radioactive material will be eventually reduced to just a few atoms, say, eight. In approximately one more half life, half of these will disappear leaving only four, then half of these leaving two, then one, and finally none.

While one cannot say exactly when the last atom of any given radioactive sample will disappear, it is easy to calculate the probability that all will have decayed at any future time. For more and more remote dates, the odds approach one to nothing. For example, it can be proved that there is a 99% chance that all of a radioactive sample will have disappeared by a time t:

$$t = (58.37 + 3.322 A T_{1/2}) T_{1/2}$$

The relationship between the decay constant and the half life is simply derived:

$$A(T_{1/2}) = \frac{1}{2} A(0) = A(0) e^{-\lambda T_{1/2}}$$

Whence:

$$\frac{1}{2} = e^{-\lambda T_{1/2}}$$

Taking the natural logarithm of each side:

$$-\ln 2 = -\lambda T_{1/2}$$

Since:

$$\ln 2 = 0.693\ldots,$$

$$T_{1/2} = \frac{0.693}{\lambda} : \quad \text{or}$$

$$\lambda = \frac{0.693}{T_{1/2}}$$

Since one is much more likely to know $T_{1/2}$ than λ, it is convenient to rewrite the decay laws in terms of the half life:

$$\Delta A = -\left(\frac{0.693}{T_{1/2}}\right) N \Delta t$$

$$A(t) = A(0) e^{-0.693 \left(\frac{t}{T_{1/2}}\right)}$$

Another quantity that is sometimes of interest is the average life of any radioactive nucleus. One might guess that the average life must be equal to the half life but that is not true. Because a few atoms survive for very long times, the average life tends to be higher. In fact:

$$T_{AVE} = \frac{1}{\lambda} \doteq 1.44\, T_{1/2}$$

PROBLEMS

1. Calculate the decay constant of H-3 from its half life.

2. What is the average life of a C-14 nucleus?

3. What is the activity of one kilogram of U-238?

4. A patient is given one millicurie of colloidal Au-198. If none is excreted, how many days must pass before the body burden is reduced to one microcurie?

5. The earth is said to be five billion years old. If that is true, what fraction of the original supply of U-238 now remains?

6. A board found in an ancient tomb contains only 60% as much C-14 as fresh-cut lumber. How old is the tomb?

Chapter X NUCLEAR DECAY

Perhaps the most immediately obvious fact of nuclear physics is that only a very few combinations of N and Z are possible for stable nuclei. If one were to choose random pairs of values, he would discover that in most cases the particles could not be made to stick together at all; such would be true for, say, $Z = 10$ and $N = 35$. Only about 280 combinations will yield stable nuclei; about a thousand more, nuclei of varying degrees of instability.

If we examine the N/Z values of stable nuclei, we find that:
a. Z can not exceed 82,
b. N is equal to or a little larger than Z, H-1 and He-3 being the only exceptions,
c. Nuclei with N and Z both even are by far the most common; those with one or the other odd are in the minority. H-2, Li-6, B-10, and N-14 are the only stable nuclei with N and Z both odd.

As a rule, nuclei with combinations of N and Z nearly equal to those of stable nuclei, though subject to decay, are comparatively long-lived. Progressively more remote combinations tend to yield nuclei that have shorter and shorter half lives.

ALPHA EMISSION

As mentioned earlier, one cause of nuclear instability is a large quantity of positive charge. It would seem that the nucleus could relieve that condition by emitting protons; however such rarely occurs. Instead, charge is reduced by expelling an *alpha particle,* a cluster of two protons and two neutrons identical to the nucleus of a helium atom.

The reason for this behavior is suggested by the orbital diagrams previously discussed. These show that the nucleons are grouped together by fours, half protons and half neutrons. If, as the diagram implies, the members of each group are tightly bound, then it is plausible that such a group would be ejected as a unit.

Every nucleus heavier than Pb-208 is subject to alpha decay. However, if a particular nuclide is also liable to some other mode of disintegration as well, say, beta emission, then its alpha instability may not be observed.

Heavy nuclides give rise to long series of radioactive daughters. Uranium, for example, with $Z = 92$ must give off at least five alphas to reduce the number of protons to 82. Actually even more are emitted since beta processes along the way convert neutrons into protons.

Alpha decay reduces N and Z by two units and A by four. For example:

$$^{238}_{92}U \rightarrow {}^{236}_{90}Th + {}^{4}_{2}\alpha \ ; \quad {}^{226}_{88}Ra \rightarrow {}^{224}_{86}Rn + {}^{4}_{2}\alpha$$

In general the alpha decay formula can be written:

$$^{A}_{Z}X \rightarrow {}^{(A-4)}_{(Z-2)}Y + {}^{4}_{2}\alpha$$

Where X and Y represent the chemical symbols of the parent and daughter nuclides respectively.

Medically speaking, the most important alpha decay series is the one beginning with U-238. Not only is this series comparatively abundant, it also includes the uniquely valuable and hazardous nuclide Ra-226.

This series is abundant because the parent, U-238, has a half life of nearly five billion years, the age of the earth. Consequently about half of the original supply is still present.

For practical reasons, this series can be divided into two portions, that which precedes Ra-226, and that which, beginning with Ra-226 finally ends with stable Pb-206. The first portion of the series is:

Nuclear Decay

$$^{238}_{92}\text{U} \xrightarrow[4.5 \times 10^9 \text{ yr}]{\alpha} {}^{234}_{90}\text{Th} \xrightarrow[24 \text{ d}]{\beta} {}^{234m}_{91}\text{Pa} \xrightarrow[68 \text{ s}]{\beta} $$

$$^{234}_{92}\text{U} \xrightarrow[247{,}000 \text{ y}]{\alpha} {}^{230}_{90}\text{Th} \xrightarrow[80{,}000 \text{ y}]{\alpha} {}^{226}_{88}\text{Ra} \xrightarrow[1600 \text{ y}]{} $$

The nuclides between U-238 and R-226 are of little practical importance.

Notice that two of the nuclides in this series emit beta rays. This fact does not mean that they are stable against alpha decay, but only that the N/Z imbalance is so great that beta decay takes precedence.

The second half of the series is:

$$^{226}_{88}\text{Ra} \xrightarrow[1600 \text{ y}]{\alpha} {}^{222}_{86}\text{Rn} \xrightarrow[3.8 \text{ m}]{\alpha} {}^{218}_{84}\text{Po} \xrightarrow[3 \text{ m}]{\alpha} $$

$$^{214}_{82}\text{Pb} \xrightarrow[27 \text{ m}]{\beta} {}^{214}_{83}\text{Bi} \xrightarrow[20 \text{ min}]{\beta} {}^{214}_{84}\text{Po} \xrightarrow[\sim\phi]{\alpha} $$

$$^{210}_{82}\text{Pb} \xrightarrow[20 \text{ y}]{\beta} {}^{210}_{83}\text{Bi} \xrightarrow[5 \text{ d}]{\beta} {}^{210}_{84}\text{Po} \xrightarrow[138 \text{ d}]{\alpha} $$

$^{206}_{82}\text{Pb}$ (stable)

The series is actually somewhat more complicated than shown above since some of the nuclides undergo either alpha or beta decay. Only the predominant mode has been shown. However the end result is always Pb-206.

Radium excepted, none of these nuclides is of much importance in itself. All but Pb-210 and Po-210 have very short half lives. Pb-210, with a half life of 20 years, finds limited commercial application. Po-210 also has a few uses. The final decay product, Pb-206, constitutes about 24% of the total quantity of lead on earth.

The medical usefulness of radium comes about this way: Suppose that one milligram (that is, one millicurie very nearly) of radium is sealed in a leakproof container. Radon gas will begin to collect so that within a month or so one millicurie of it will also be present. Thereafter the gas will decay as rapidly as it is formed so that the total quantity will be constant from then on. One millicurie amounts of the next four nuclides will also build up at the same time. The fifth nuclide Pb-210 will also start to accumulate, but because of its long half life, over a century would pass before its activity reaches the equilibrium value of one millicurie. Hence the container, which originally held only one milligram of radium, will soon have one millicurie each of Rn-222, Po-218, Pb-214, and Po-214 within it.

Several of these nuclides are formed with excited nuclei. In going to the ground state, they will emit gamma rays, which—unless the walls of the container are exceedingly thick—will escape to the outside. In other words, the container has become a source of gamma rays.

For medical use hollow needles and tubes of platinum are filled with a radium salt. Nowadays the sulfate is used although the chloride was common in the past. After being aged long enough for equilibrium quantities of the first four daughter products to build up, these are effective, stable sources of gamma rays. Their clinical advantage is that one can often give a sufficiently high dose to cancerous tissue without overexposing adjacent healthy tissue in situations where that may be impossible with external therapy machines.

Curie for curie, Ra-226 is one of the most hazardous of all known radionuclides. Here are the reasons:

1) Radium is chemically similar to calcium. Hence if it ever enters the body, it will be deposited in bone and once there it will stay for a long time.
2) Because of its long half life, it remains at full strength for the lifetime of the individual concerned.
3) Radium emits alpha particles which are roughly twenty times more damaging than equal doses of x- or beta rays.

According to current recommendations, the maximum permissible body burden of Ra-226 is a tenth of a microgram (or microcurie). This is an extremely small amount; a single ten mg medical source could theoretically bring 100,000 people up to their maximum permissible body burden. Clearly the breakage of such a source is a disaster of the first magnitude.

BETA EMISSION

Beta decay can be easily understood by means of the orbital diagrams introduced earlier. Consider the nuclide B-13. Its

Nuclear Decay

ground state configuration is as shown in Figure 18a. Notice that the column of neutrons is somewhat higher than that of the protons. Clearly the energy of the system could be reduced if one neutron were placed over beside the uppermost proton as shown in Figure 18b. Such is impossible, of course, since that position is reserved for protons. However if the neutron first decays into a proton as described in Chapter V the change of position can and does occur.

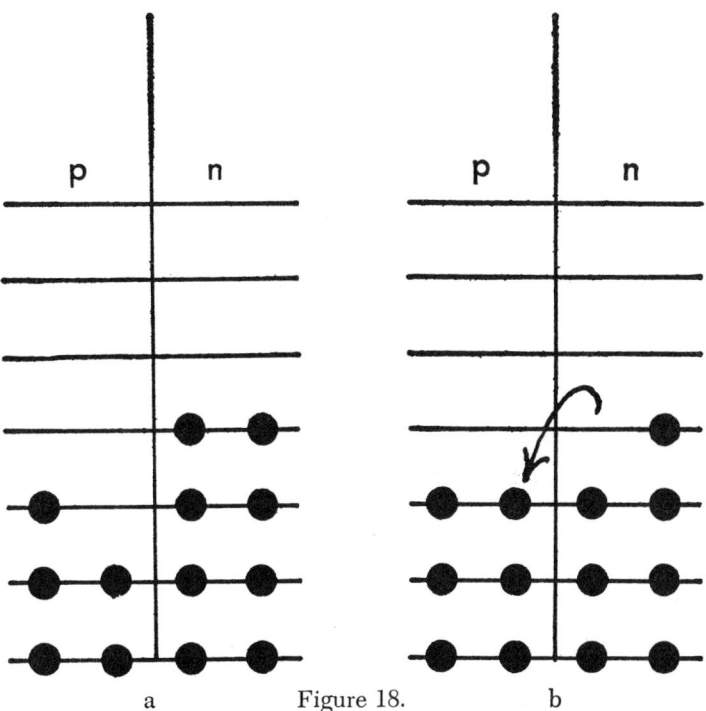

Figure 18.

Immediately after disintegration, the electron and the antineutrino leave the nucleus. The antineutrino leaves because there is no force to restrain it; the electron, although it is strongly attracted by the surrounding protons, can not be confined to a region as small as an atomic nucleus with the available binding energy.

The newly formed proton drops into the lowest available energy level as shown in Figure 18c. The result of this process has been to increase Z from five to six and to reduce N from

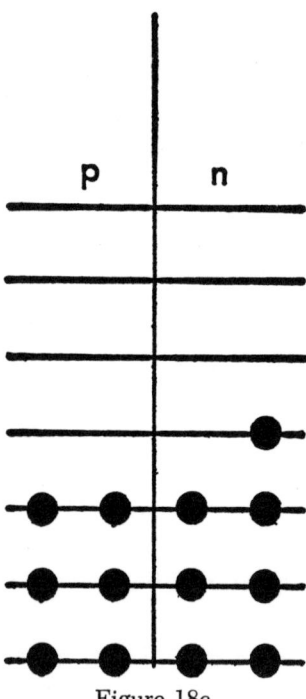

Figure 18c.

eight to seven. The nucleus now belongs to the element carbon instead of boron.

The reaction is conventionally expressed:

$$^{13}_{5}B_8 \longrightarrow {}^{13}_{6}C_7 + e^- + \tilde{\nu} + \text{energy}$$

In this particular reaction, the released energy appears as kinetic energy of the electron and the antineutrino. In many beta-active nuclei, especially those with larger Z values, some of the energy is given off as gamma rays.

What happens to the B-13 nucleus happens to every nucleus for which the column of neutrons in the orbital diagram is appreciably higher than the column of protons. Such nuclei are said to be *neutron-rich*, or what amounts to the same thing, *proton-poor*. They will undergo as many successive disintegrations as necessary to equalize the heights of the two columns.

There are also nuclei that are proton-rich. Consider N-12 whose nucleonic configuration is shown in Figure 19a. A more stable arrangement would be achieved if the proton were converted into a neutron and transferred to the other side as shown in Figure 19b.

To accomplish this transformation, the proton may combine with an orbital electron, or—if sufficient energy is available—turn directly into a neutron, positron, and a neutrino.

In either case the nucleus loses a proton and gains a neutron so that Z decreases by one, N increases by one, and A stays the same. As in the case of neutron-rich nuclides, parent and daughter are isobars.

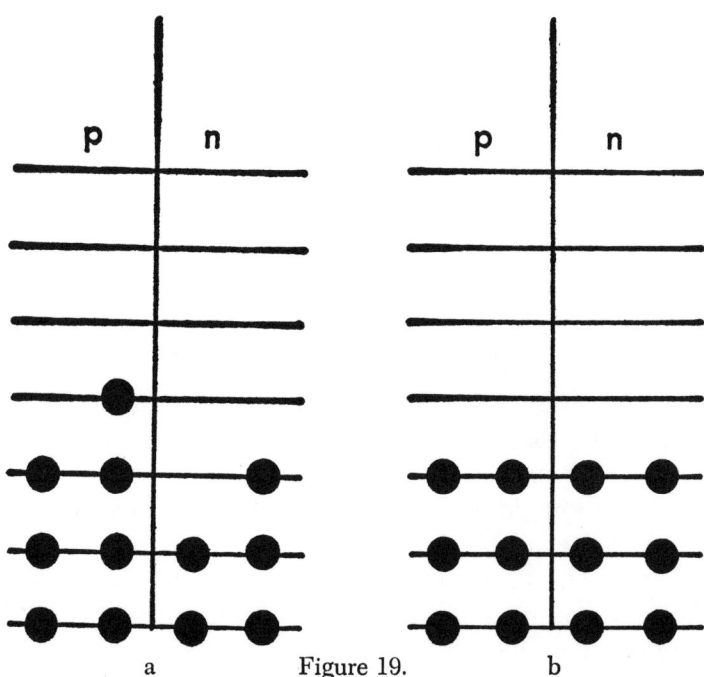

Figure 19.

Our present example, N-12, is quite unstable, so that more than enough energy is available for direct proton decay. The reaction is:

$$^{12}_{7}N_5 \longrightarrow {}^{12}_{6}C_6 + e^+ + \nu + \text{energy}$$

The general decay pattern for positron emission is:

$$^{A}_{Z}X \longrightarrow {}^{A}_{Z-1}Y + e^+ + \nu + \text{energy}$$

As in the case of beta-minus decay, the energy is released as kinetic energy of the positron and the neutrino. If the daughter is formed in an excited state, gamma rays are also given off.

Electron capture could occur, of course, but only when an orbital electron, usually one from the K shell, happens to come near the nucleus. As this is a fairly rare event compared to other nuclear processes, nuclei that can decay by positron emission usually do so long before an electron comes by.

However a number of proton-rich nuclei do not have sufficient energy for positron emission. For example:

$$^{7}_{4}Be + e \longrightarrow ^{7}_{3}Li + \nu$$

The general decay pattern is:

$$^{A}_{Z}X + e^- \longrightarrow ^{A}_{Z-1}Y + \nu + \text{energy}$$

Some nuclides with barely sufficient energy for positron emission may go either way:

$$^{48}_{23}V \longrightarrow ^{48}_{22}Ti + \beta^+ + \nu \quad (49\%)$$

$$^{48}_{23}V + e^- \longrightarrow ^{48}_{22}T + \nu \quad (51\%)$$

Expressed in mass units, the energy available for positron emission is the mass difference of the parent and daughter nuclei. This must provide both the rest mass energy of the positron and the kinetic energy (if any) of the final products.

The energy available for electron capture is the mass difference of the parent and the daughter nuclei plus also the total mass of the captured electron. The only energy output is the kinetic energy of the neutrino which may have any value from zero up.

The above statements presuppose that the daughter is formed in the ground state. If it is not, then there must be sufficient initial energy to form the emitted photons as well.

Consider two neighboring isobars $^{A}_{Z}X$ and $^{A}_{Z-1}Y$:

1. If $M(A, Z) < M(A, Z-1) - m$, then $^{A}_{Z}X$ is either stable or a negatron emitter.
2. If $M(A, Z-1) - m \leq M(A, Z) < M(A, Z-1) + m$, then $^{A}_{Z}X$ is subject to electron capture only.

3. If $M(A, Z) \geq M(A, Z-1) + m$
then ^A_ZX is subject to both electron capture and positron emission. In most cases the latter mode overwhelms the former.

BETA DECAY—THE LIGHT ELEMENTS

In the light elements, namely those with Z not greater than twenty, the electrostatic repulsion is comparatively weak. As a result, the proton and the neutron levels are of about equal height. Since for a stable nucleus the diagram must be filled equally on both sides, the number of protons and neutrons must obviously be about equal. In fact, stability is possible in this region only if $N = Z - 1$, Z, $Z + 1$, or $Z + 2$. This condition is not sufficient however, as there are several nuclides in this region fulfilling it that are nevertheless radioactive.

There are only two stable nuclides for which $N = Z - 1$, namely H-1 and He-3. Otherwise it can be said that a stable nucleus must have at least as many neutrons as protons.

Those nuclides with N equal to Z are of special interest. Most of them are stable and abundant. Be-8 however does not exist because its nuclei quickly fall apart into pairs of alpha particles. In this group of nuclides are four that violate the general rule that *nuclides are unstable if N and Z are both odd;* namely H-2, Li-6, B-10, and N-14. These are the lightest possible odd-odd combinations, and their anomolous stability is due to special causes.

Among the light nuclei, all with $N = Z + 1$ are beta stable except A-37 which decays to Cl-37. Here is the first evidence of the disruptive power of the electrostatic force. The nucleus of He-5, although presumably stable against beta decay, disintegrates instantly into an alpha particle and a proton.

Of the light nuclides with $N = Z + 2$, about half are stable and half not.

Beyond $Z = 20$, the repulsive force between protons is great enough to force the proton levels significantly higher than the corresponding ones for neutrons. Accordingly, if the heights of the proton and neutron columns are to be equal, the number of neutrons must exceed that of the protons. As one goes to higher and higher Z values, the N/Z ratio increases as shown in Figure 20.

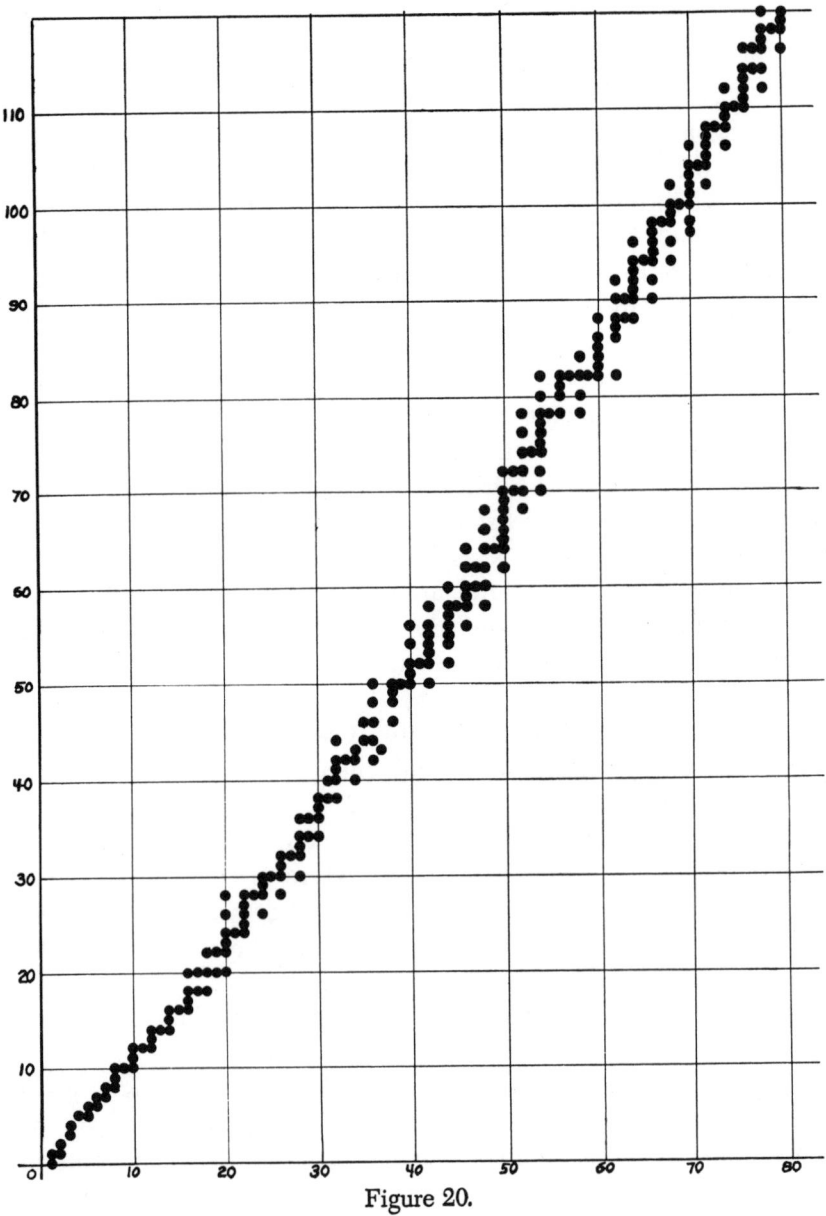

Figure 20.

Nuclear Decay

There are several points in the diagram for most elements, one for each stable isotope. Two elements, however, *promethium* and *technecium,* have no stable isotopes at all. The designated nuclides beyond lead are beta-stable, but as we have seen, are subject to alpha decay and sometimes fission.

BETA DECAY AND ISOBARIC FAMILIES

When a nucleus undergoes beta decay, the total number of nucleons in it does not change since a proton turns into a neutron or vice versa. Hence, the total count $N + Z$ remains constant. In other words, the parent and the daughter nuclei are isobars each of the other. Such being true, study of isobaric families is important.

For any given value of A, there could conceivably be $A + 1$ isobars, with values of Z from zero to A. Thus for $A = 6$ one might have:

$${}^{6}_{0}?\quad {}^{6}_{1}H\quad {}^{6}_{2}He\quad {}^{6}_{3}Li\quad {}^{6}_{4}Be\quad {}^{6}_{5}B\quad {}^{6}_{6}C$$

Actually most of these combinations cannot be assembled even for a moment. Suppose one could press six protons close enough together to form a nucleus of C-6. The electrostatic repulsion would immediately blow them apart. Of the nuclides suggested, only He-6, Li-6, and Be-6 can be formed at all. Li-6 is stable and the other two have half lives of a little less than a second.

Two types of isobaric families must be distinguished; those for which A is even and those for which A is odd.

In isobaric families of odd A, only one nuclide can be stable. All the others will progressively decay until the stable isobar is reached. Consider the family $A = 99$:

$${}^{99}_{41}Nb \longrightarrow {}^{99}_{42}Mo \longrightarrow {}^{99}_{43}Tc \longrightarrow {}^{99}_{44}Ru \longleftarrow {}^{99}_{45}Rh \longleftarrow {}^{99}_{46}Pd \longleftarrow {}^{99}_{47}Ag$$
$$2.4\ m \qquad 67\ h \qquad 2 \times 10^5\ y \qquad 16\ d \qquad 22\ m \qquad \sim\phi s$$

A nucleus of Nb-99 will under go three successive decays finally becoming a stable nucleus of Ru-99. At each step a surplus neutron is converted into a proton, and an electron and an antineutrino are emitted. On the other side, a nucleus of Ag-99 decays in three steps, also finally arriving at Ru-99.

In the family above, the half lives of the various members are progressively shorter the further they are from the stable member. This is the usual trend although exceptions occur in certain other isobaric families. For example:

$$^{127}_{51}\text{Sb} \xrightarrow[93 \text{ nr}]{} {}^{127}_{52}\text{Te} \xrightarrow[9.4 \text{ hr}]{} {}^{127}_{53}\text{I (stable)}$$

$$^{62}_{30}\text{Zn} \xrightarrow[9.1 \text{ h}]{} {}^{62}_{29}\text{Cu} \xrightarrow[9.8 \text{ m}]{} {}^{62}_{28}\text{Ni (stable)}$$

The reasons for these anomalies are known, but are rather complicated, and for our purposes, unimportant.

Whereas isobaric families with odd A can have only one stable member, those of even A often have two. This phenomenon results from the strong tendency of like nucleons to pair up. As the orbital diagram shows, there is room for just two protons or two neutrons in each level. If only one is present, it will either try to escape or attract another to join it. In nuclides of odd A, there must be either an odd proton or an odd neutron; it doesn't too much matter which. Among the stable odd-A nuclides, there are about as many with odd Z as with odd N.

If A is even however, then N and Z must either be both even or both odd. If N and Z are even, there are no unpaired nucleons at all; if N and Z are both odd, there is one of each. This unfavorable situation can be relieved if the proton becomes a neutron or vice versa. So strong is this tendency that no stable odd-odd nuclides exist except the first four previously mentioned. In them special conditions prevail.

In the isobaric family with even A, the most stable position might be occupied by either an even-even or an odd-odd nuclide. If the nuclide in this position is even-even, then it is the only stable member. All the proton-poor nuclides decay by successive negatron emission and all the proton-rich nuclides decay by either positron emission or electron capture. For example if A equals 32:

$$^{32}_{14}\text{Si} \xrightarrow[700 \text{ y}]{} {}^{32}_{15}\text{P} \xrightarrow[14.3 \text{ d}]{} {}^{32}_{16}\text{S} \xleftarrow[0.30 \text{ s}]{} {}^{32}_{17}\text{Cl} \xleftarrow[?]{} {}^{32}_{18}\text{A}$$

All members eventually become S-32. Notice the long half life of Si-32. Though further from the central position than P-32, it is very reluctant to change its status from even-even to odd-odd.

It frequently occurs that the most stable position in the family is filled by an odd-odd nuclide. Since it can not itself be stable, the two adjacent even-even nuclides must be. Such a sequence can be divided into five parts:

neutron-rich nuclides	low-Z stable nuclide	central odd-odd nuclide	high-Z stable nuclide	proton rich nuclides

The neutron-rich nuclides decay to the low-Z stable member and the proton-rich nuclides decay to the high-Z stable member. The central nuclide will decay to the least energetic of its neighbors as a rule. For certain even values of A, decay of the central member in either direction is about equally attractive; hence it will decay both by negatron emission as well as positron emission and/or electron capture. Here are some examples:

$$^{50}_{21}\text{Sc} \xrightarrow[1.7 \text{ m}]{\beta^-} {}^{50}_{22}\text{Ti} \xleftarrow[6 \times 10^{15} \text{ y}]{ec} {}^{50}_{23}\text{V} \quad {}^{50}_{24}\text{Cr} \xleftarrow[2 \text{ m}]{\beta^+} {}^{50}_{25}\text{Mn}$$

$$^{82}_{33}\text{As} \xrightarrow{\beta^-} {}^{82}_{34}\text{Se} \quad {}^{82}_{35}\text{Br} \xrightarrow[35 \text{ n}]{\beta^-} {}^{82}_{36}\text{Kr} \xleftarrow[1.2 \text{ m}]{\beta^+} {}^{82}\text{Rb} \xleftarrow[25 \text{ d}]{ec} {}^{82}_{38}\text{Sr}$$

$$^{108}_{45}\text{Rh} \xrightarrow[17 \text{ s}]{\beta^-} {}^{108}_{45}\text{Pd} \xleftarrow[2.4 \text{ m}]{ec, \beta^+} {}^{108}_{47}\text{Ag} \xrightarrow{\beta^-} {}^{108}_{48}\text{Cd} \xleftarrow[39 \text{ m}]{\beta^+} {}^{108}_{49}\text{In}$$

When a very light nucleus undergoes beta decay, the transformed nucleon usually goes directly to the lowest (which is often the only) available position. The energy released appears as kinetic energy of the electron and the neutrino. For a given pair of parent-daughter nuclides, the total energy per disintegration is constant, but it may be divided in any proportion between the electron and the neutrino. On the average, about one third to one half of the energy is carried away by the electron.

96   Basic Nuclear Physics for Medical Personnel

In larger nuclei the situation is not so simple. Such nuclei have a number of excited states within a few MeV of the ground state, and the daughter nucleus is often created in one of these. Whenever that happens, one or more gamma rays are usually released shortly after the beta transformation.

Figure 21 shows the energy level diagrams for Co-60 and its daughter, Ni-60. In this reaction the daughter nucleus is nearly always formed in the fourth excited state. This generally decays to the first excited state by emitting a gamma ray of 1.17 MeV. A further de-excitation with emission of a 1.33 MeV

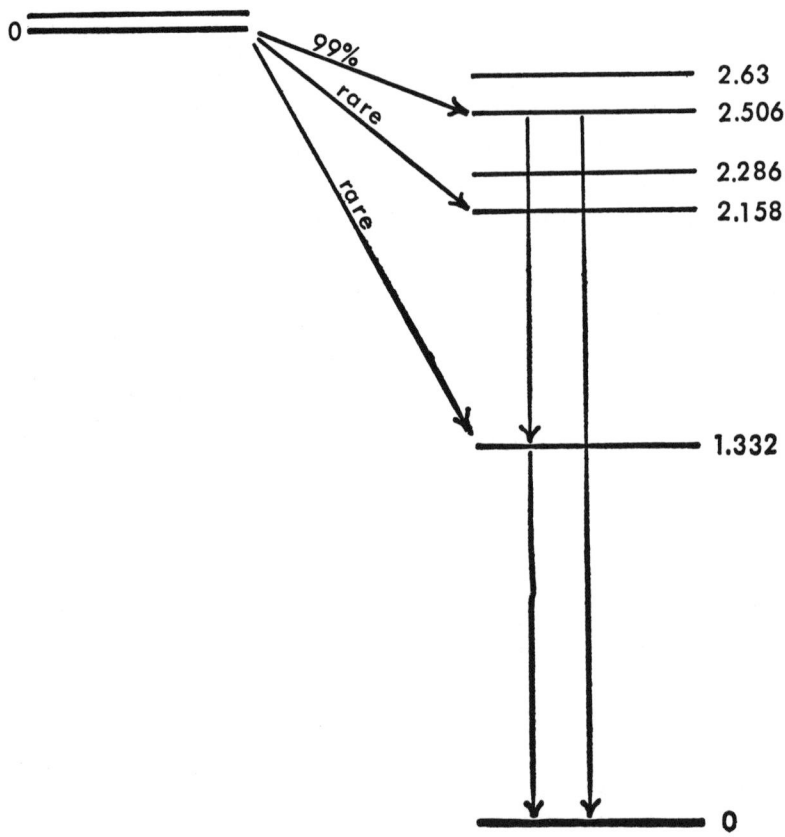

Figure 21.

gamma ray brings the nucleus to its ground state. Very rarely the fourth excited state may decay directly to the ground state by ejecting a 2.50 MeV photon. This direct transition occurs so seldom because it forces the nucleus to make a large and sudden change in its rate of spin.

CONVERSION ELECTRONS

Sometimes the surplus energy, instead of being organized into a photon, is deposited directly onto an atomic electron. K electrons, being nearest the nucleus, are usually the ones affected. This electron—called a conversion electron—immediately shoots out of the atom at high speed. Its kinetic energy is equal to the energy released in the nucleus minus the binding energy of the electron in its shell. Internal conversion is most likely for high values of Z and low values of energy.

ISOMERS

Isomers are most commonly produced by beta decay. For example, when Mo-99 decays to Tc-99, the latter is usually produced in its second excited state which is metastable with a half life of six hours.

Metastable states decay according to the radioactive decay law; that is, one half of the existing states disappear in equal increments of time.

Isomers are of especial interest in medicine because they are pure gamma emitters. Whenever a radioactive drug is introduced into the human body, all its radiation (except the neutrinos) cause damage. If the drug used is for diagnostic purposes, it is usually only the gamma rays that are practically detectable. Alpha and beta particles harm the patient without contributing anything useful. Consequently, isomers, especially Tc-99m, are popular for diagnostic procedures.

Chapter XI NATURAL RADIOACTIVITY

THE RADIOACTIVE NUCLIDES that now occur naturally upon the earth may be divided into three groups:

1. Original radionuclides
2. Radioactive descendents of the original radionuclides
3. Radionuclides produced by natural agents

ORIGINAL RADIONUCLIDES

The nuclides that compose the sun and its planets including the earth, came into being about five billion years ago. There are several theories as to just what happened at that time, none definitely proved. Besides the stable nuclides, hundreds of unstable ones also came into being as well, most of which vanished long ago. Several however had half lives so long that the original supply is not yet exhausted. These are listed in Table II.

Most of them are rare curiosities. K-40 is important because it is rather abundant and contributes significantly to biological exposure. Ca-48, although abundant and an essential element of the body, has such a long half life that its activity is almost immeasurably small. The heavy nuclides Th-232, U-235, and U-238 are not only abundant but beget long series of highly active, useful, and dangerous descendents.

There are three distinct kinds of original radionuclides:

1. Those with a neutron-proton imbalance. These decay by beta emission or by electron capture. Of these K-40 is by far the most important.

2. The light alpha emitters. For values of Z between 60 and

TABLE II

Z	Nuclide	$T_{1/2}$ (yr)	Decay	Products:
19	K-40	1.3×10^9	β^-, ec	Ca-40, A-40
20	Ca-48	2×10^{16}	β^-	Sc-48 $\to$ Ti-48
23	V-50	6×10^{14}	β^- & ec	Ti-50, Cr-50
37	Rb-87	4.7×10^{10}	β^-	Sr-87
41	Nb-91	long	ec	Zr-91
57	La-138	1.1×10^{11}	ec, β^+	Ba-138
58	Ce-142	5×10^{15}	α	Ba-138
60	Nd-144	5×10^{15}	α	Ce-140
62	Sm-147	1.06×10^{11}	α	Nd-143
62	Sm-148	1.2×10^{13}	α	Nd-144 $\to$ Ce-140
62	Sm-149	4×10^{14}	α	Nd-145
64	Gd-152	1.1×10^{14}	α	Sm-148 $\to$ Nd-144 $\to$ Ce-140
71	Lu-176	2.1×10^{10}	β^-	Hf-176
72	Hf-174	4.3×10^{15}	α	Yb-170
75	Re-187	7×10^{10}	β^-	Os-187
78	Pt-190	7×10^{11}	α	Os-186
78	Pt-192	10^{15}	α	Os-188
82	Pb-204	1.4×10^{19}	α	Hg-200
83	Bi-209	2×10^{17}	α	Tl-207 $\to$ Bb-207
90	Th-232	1.39×10^{10}	α, SF	series to Pb-208
92	U-235	7.13×10^8	α, SF	series to Pb-207
92	U-238	4.51×10^9	α, SF	series to Pb-206

83, there are a number of nuclides which, although the neutron-proton ratio is correct, have weak internal structure. The situation is similar to that often found in chemistry, where certain combinations of atoms cannot form tight bonds although the valence requirements are met. The resulting compounds are unstable and decompose either spontaneously or under the slightest provocation.

Since the nucleons are weakly held in such nuclei, the protonic repulsion forces alpha emission. The process goes on very slowly however because the electrostatic force is not very strong.

3. The heavy alpha emitters. For Z values above 83, no nuclear structure can withstand the electrostatic repulsion. Nearly all nuclides in this region decay very rapidly. Only thorium-232, uranium-235, and uranium-238 have half lives long enough to have survived from the time of creation.

These two radio-elements are fairly abundant and may be regarded as stable for many applications. Quite apart from their radioactivity both have valuable physical and chemical properties. Uranium is used as a yellow pigment in ceramics and glass; it has some value as a radiation shield because of its high density and high atomic number.

Thorium oxide greatly improves the emissivity of vacuum tube cathodes and is an essential ingredient of efficient gas mantles. The physical properties of metallic magnesium are improved when it is alloyed with thorium. Compounds of this latter element were once used as contrast media in diagnostic x-ray examinations but have been discontinued, because of the traces of radioactivity remaining in the body are thought to be unnecessarily hazardous. Anyway, compounds of iodine and barium are quite satisfactory.

Both thorium and uranium are alpha emitters. If these elements are separated from their daughter products, they give off but few gamma rays and no beta particles. As their alpha radiation is absorbed in a few centimeters of air and in exceedingly thin layers of solids or liquids, either element can be handled without significant radiation exposure. One should, however, be especially careful not to ingest compounds of either element.

RADIOACTIVE DESCENDENTS OF ORIGINAL RADIONUCLIDES

Most of the nuclides listed in Table II decay directly to a stable nuclide. A few decay to another radionuclide which then decays to a stable form. The three nuclides Th-232, U-235, and U-238, however, originate long series of radioactive nuclides. Each member of such a series decays to the next until finally it becomes a stable isotope of lead. The series beginning with U-238 has been discussed in the previous chapter; the others are of little medical interest.

RADIONUCLIDES PRODUCED BY NATURE

There are many of these in trace amount, but the only one of much practical importance is C-14. It is produced by cosmic rays striking the atmosphere. Newly formed atoms of C-14 combine with oxygen to form carbon dioxide. This is inhaled by plants and incorporated into their tissues. Since animals feed directly or indirectly on plants, their bodies also contain C-14 in about the same relative abundance as the air.

Once the organism dies, the intake of C-14 ceases. Thereafter, that present decays so that old lumber, fabric, and other organic residues are poorer in this isotope than fresh materials. This fact makes it possible (if not easy) to determine the approximate age of any such substance by assaying its C-14 content. The half life of this nuclide (5700 yr) is such as to make it especially valuable for ages of 500, to 5,000 years, just the time span of greatest interest to historians.

Chapter XII ARTIFICIAL RADIOACTIVITY

As we have seen, only a few of the many possible radionuclides occur naturally upon the earth, and most of these are of little medical, scientific, or industrial use. Were we dependent solely upon natural radioactivity, neither the wonders nor the dangers of the *nuclear age* would exist. Most of the radionuclides now employed in medicine, industry, and research have come into being through human effort.

There are three important methods of producing radionuclides artificially:

1. by nuclear fission,
2. by neutron capture, and
3. by charged particle irradiation

Of these, the last-named is the oldest. In 1934 the Joliot-Curies bombarded aluminum-27 with alpha particles and produced phosphorus-30—admittedly in quantities much too small to be detected chemically. Although the production of radionuclides by charged particle interactions has been long known and often leads to highly desirable end products, it is slow, poor in yield, and expensive. As a result radionuclides which must be thus manufactured enjoy but limited use.

Fission products, on the contrary, are in embarrassing over supply as they are—so to speak—the ashes of a nuclear reactor. In fact, one of the pressing problems of nuclear technology is their safe disposal. Unfortunately only a few of these nuclides are useful. Cesium-137 was disappointing as source material for radiation therapy units; the much more expensive cobalt-60 is so decisively superior that it has almost completely eliminated its rival. Strontium-90 has been made up into beta-ray applicators, but the need for such devices is minimal. Of the many fission products only iodine-131 is of particular medical consequence.

Most radionuclides used in appreciable quantities are manufactured by exposing (usually) stable nuclides to the intense flux of neutrons found in the interior of a nuclear reactor. Cobalt-60, gold-198, and molybdenum-99 are just a few of the nuclides thus produced. Regrettably, all such nuclides are negative electron emitters. Nuclides which emit positrons, undergo electron capture, or have useful isomeric states cannot be directly produced by neutron capture.

NUCLEAR FISSION

Nuclear fission comes about through the interplay of two types of force that are always present in the atomic nucleus: the nuclear force which holds it together and the electrostatic repulsion which tends to blow it apart. As we have seen, the latter becomes relatively stronger with increasing atomic number.

No nucleus with more than eighty-two protons can stick together indefinitely. Sooner or later it will reduce its size either by ejecting an alpha particle or by splitting apart. This latter process is called *fission*.

Fission is of two kinds, *spontaneous* and *induced*. The first, as the name implies, needs no outside encouragement. As is true of other forms of radioactive disintegration, spontaneous fission follows the exponential decay law. Among the naturally occurring nuclides, spontaneous fission is exceedingly rare. (Obviously. Those nuclides that fission readily have long since vanished.) In thorium-232 only one nucleus undergoes fission for every hundred billion that decay by alpha emission. In uranium-235 the ratio is about one in a hundred million and in uranium-238 about one in a million. Among the artificially produced transuranic nuclides fission may be the dominant mode of decay. In curium-248 about eleven per cent of the nuclei fission spontaneously, while in californium-254 nearly all do so.

By induced fission we mean fission which occurs in response to an external stimulus. A number of high-Z nuclei can be made to split apart by irradiating them with neutrons. Graphically speaking, what happens is this: Those nuclei which can be induced to fission are already *on the verge* as proved by the fact that a tiny fraction of them fission spontaneously. Whenever a neutron is absorbed, the daughter nucleus is initially excited by six to ten MeV above its ground state. Although thrown into violent agitation lighter nuclei will soon eject the excess energy by gamma radiation. Those that are large and structurally weak however may shake themselves apart. In uranium-235 and plutonium-239 the likelihood of an absorbed neutron triggering off fission is about seventy-five per cent.

A particularly curious fact about fission is that the nucleus almost never splits into equal halves. Instead, one fragment is usually about 1.6 times heavier than the other. Thus when uranium-235 decays by fission, there result abundant quantities of nuclides with masses between 90 and 100 AMU and also between 135 and 145 AMU. The yield of nuclides with masses in the vicinity of $235/2 = 117$ AMU is low. This behavior is contrary to both common sense and the prediction of simple nuclear theory.

A serious consequence of this lopsided division is that strontium-90 and cesium-137—both with half lives of about thirty years—are among the principal fission products. Whereas most of the others decay within a reasonable period of time, these two endure for centuries. Providing safe, long-term storage for them is exceedingly difficult, and occasional leakage or spills are hardly to be avoided altogether.

As we have seen, the relative number of neutrons required for nuclear stability becomes ever greater as Z increases. For the light elements such as carbon-12 and oxygen-16, maximum stability is achieved when $N = Z$ that is when protons and neutrons are present in equal numbers.

For elements at the end of the periodic table (thorium-232, uranium-235, and uranium-238) beta stability demands an N/Z ratio approximately equal to 1.57.

At $A = 95$ (the average atomic mass of the lighter fission fragments) the optimum value for N/Z is 1.26 and at $A = 140$ (the average atomic mass of the heavier fission fragments) the optimum value is 1.41.

Since the N/Z ratio of the initial fission fragments is necessarily that of the original nuclide—namely 1.57—obviously these fragments have far too many neutrons to be stable. The imbalance is so great that the immediate daughter nuclei sometimes emit neutrons without waiting for beta decay to rectify matters.

On the average each fission results in the emission of about three neutrons. It is these that make the chain reaction possible. If at least one of the released neutrons induces fission in another nucleus, the process will continue until the quantity of fuel is depleted (as in a reactor) or the assembly is destroyed (e.g. an atomic bomb).

To start and maintain a chain reaction, "all" one need do is bring together a sufficient quantity of fissionable material (uranium-235 or plutonium-239) and arrange it into a shape such that few neutrons can escape. A successful bomb must contain a mechanism that quickly assembles the nuclear charge into the right configuration before the explosion has begun. If the assembly time is too long, the chain reaction will begin prematurely and the fuel will be dispersed before the reaction has run to completion. The result will be a misfire—a "pop" instead of a "bang." It might be remarked however that even a nuclear "pop" is awesome by conventional standards.

Controlled fission for industrial and scientific purposes takes place in reactors or piles. The principal problem with these devices is to maintain the fuel and the control rods of neutron absorbing material positioned properly so that the reaction goes at a uniform, controllable rate. If too many neutrons escape or are absorbed in non-fissionable material, the reaction will die out. If too many neutrons are absorbed by the fuel, the pile will *run wild* and destroy itself with excessive heat. The design of a reactor is such that a true, bomb-like explosion is impossible although the container may be ruptured. For that reason reactors are placed in a second container to prevent uncontrolled spread of radiocontamination.

NEUTRON CAPTURE

Whenever a neutron is moving slowly through a material medium, the chances are that it will eventually pass near an atomic nucleus and be absorbed. The result is a new nucleus containing one more neutron than before. The absorbed neutron usually drops to the lowest empty level and a gamma ray or a conversion electron carries off the excess energy. Since the number of protons present is unchanged, the new nuclide is an isotope of the old. The interaction is represented symbolically:

$$^{A}Z_N + n \to {}^{(A+1)}Z_{(N+1)} + \gamma$$

New nuclides are prepared commercially in nuclear reactors since only such devices produce enough neutrons at reasonable cost. The nuclide to be irradiated is inclosed in a special container, placed within the core of the reactor, and left as long as necessary. For some materials seconds suffice; others require years. In practice one usually begins with a stable nuclide; an exception is uranium-238 which is irradiated to yield plutonium-239.

All too often neutron capture produces nuclides that are stable. For example hydrogen-1 becomes hydrogen-2, carbon-12 becomes carbon-13, and oxygen-16 becomes oxygen-17. Such reactions have little practical interest because the stable daughter is more cheaply available elsewhere. Usually, however, the daughter—in any case richer in neutrons than the parent—is subject to beta-minus decay and hence of potential value. A few such that have found medical application are sodium-24, potassium-42, iron-59, cobalt-60, molybdenum-99, and gold-198.

Although nuclei that absorb slow neutrons usually radiate only gamma rays, there are exceptions among the light elements. When nitrogen-14 absorbs a neutron, it may promptly eject a proton:

$$^{14}_{7}N_7 + n \to {}^{14}_{6}C_8 + p$$

This reaction is a good source of the extremely valuable radioactive isotope of carbon. While this nuclide is abundant in nature, the cost of separating it from its stable isotopes is prohibitive.

The reaction:

$$^{32}_{16}S_{16} + n \to {}^{32}_{15}P_{17} + p$$

is also important as it yields phosphorus-32, a radionuclide that has been extensively used in medicine.

The initial and final nuclides of the (n,p) reaction are neighboring isobars. Symbolically:

$$^A Z_N + n \rightarrow {}^A(Z-1)_{(N+1)} + p$$

If the parent nuclide is stable (as is ordinarily the case) then the daughter must be radioactive because it is impossible for both of two neighboring isobars to be stable.

CHARGED PARTICLE REACTIONS

When nuclei are bombarded with heavy charged particles (for example, protons, deuterons, alpha particles and the like) a great many different interactions may occur depending upon the type of particle, its energy, and of course the nature of the target nucleus. If the velocity of the incoming particle is low, it will usually be repelled by the electrostatic force before it has come close enough to disrupt the target nucleus. To the observer the interaction resembles an elastic collision of two balls. Neither is altered and the total kinetic energy of the system after collision is the same as before. Practically speaking nothing of advantage has occurred.

At higher energies the electrostatic repulsion is no longer sufficient to prevent the projectile from entering the nucleus. Once the bombarding particle is inside, any reaction for which there is sufficient energy will occur at least occasionally.

The usual kind of reaction is the emission of one or more particles, the commonest being neutrons, protons, deuterons, and alpha particles. If the energy of the incoming particle is but a few MeV only one particle can be expelled; at higher energies two or more are possible.

The charged-particle interactions of practical interest are those that lead to radionuclides not otherwise available. Since neutron-rich nuclides are produced both by fission and by neutron capture, it is the neutron-poor nuclides that we seek via charged-particle bombardment. Suitable to the purpose are the (p,n), the (p,γ), the (d,n), the (d,2n) and the (d,α) reactions.

In all of these reactions the N/Z ratio of the daughter is less than that of the parent. Hence one expects the daughter to emit positrons or at least decay by electron capture. Occasionally though, the daughter will be stable as in the reaction:

$$^{19}_{9}F_{10} + d \rightarrow {}^{20}_{10}N_{10} + n$$

Very rarely it is even a negatron emitter in spite of its decreased N/Z ratio as in the interaction:

$$^{36}_{16}S_{20} + p \rightarrow {}^{36}_{17}Cl_{19} + n$$

$$^{36}Cl \rightarrow {}^{36}Ar + e^- + \tilde{\nu}$$

Among the charged particle reactions, those that are deuteron-induced have especially high yields and often lead to useful daughter products. The (d,n) reaction demands comparatively little energy; unfortunately the daughter nucleus is frequently stable. This reaction increases both the atomic number and the mass number by one unit:

$$^{A}Z_N + d \rightarrow {}^{(A+1)}(Z+1)_N + n$$

The N/Z ratio is changed but slightly which fact accounts for the frequent stable daughter nuclides. Products of especial interest obtained from this reaction are beryllium-7, carbon-11, and cobalt-57.

Although the (d,2n) reaction requires considerably more energy than the (d,n) reaction, it is justified by the many useful nuclides that it can yield. In this reaction the mass number remains unchanged, but the atomic number is increased by one unit. The N/Z ratio becomes $(N+1)/(Z-1)$:

$$^{A}Z_N + d \rightarrow {}^{A}(Z+1)_{N-1} + 2n$$

This interaction leads to the same daughter nuclide as the (p,n) reaction, but the efficiency is usually much better. As with the (p,n) reaction the daughter is necessarily radioactive if the parent is stable. Some of the nuclides prepared by this reaction are manganese-52, iron-55, cobalt-57, cobalt-58, zinc-65, arsenic-74, yttrium-88, silver-106, iodine-130 and mercury-197.

Other deuteron reactions are seldom important. For example,

the (d,α) reaction rarely does anything that cannot be done better otherwise. The following reaction is exceptional:

$$^{20}_{10}Ne_{10} + d \rightarrow ^{18}_{9}F_{9} + \alpha$$

Deuterons are fired into a gas target of neon-20. The resulting fluorine-18 immediately adheres to the side of the containing vessel from which it can be removed by suitable solvents. This nuclide is useful for diagnostic examinations of bone. As its half life is less than two hours, it must be prepared at or near the scene as needed.

Nuclides produced by charged particle bombardment are rare and costly for two reasons. First, one needs a charged particle accelerator, for example, a cyclotron. Such machines are expensive to build, to operate, and to maintain. Second, of the charged particles incident upon the target, only one in roughly ten thousand interact with a nucleus; the others are unproductively brought to rest by collisions with electrons. Thus the efficiency is less than 0.01% at best.

Chapter XIII
THE INTERACTION OF RADIATION WITH MATTER

Both the dangers and the benefits of radiation are due to its interactions with matter. Since the nature of those interactions depends upon the type of radiation, we will divide the discussion into four parts:

1. protons
2. electrons
3. neutrons
4. photons.

PROTONS

As all heavy charged particles interact with matter in much the same way, we will discuss only the proton in detail and then briefly mention some of the differences between its behavior and that of other particles such as the deuteron and alpha particle.

Protons moving rapidly through matter interact almost entirely with electrons; only one proton in about ten thousand ever encounters an atomic nucleus. When the proton passes near an electron, it exerts a strong attractive force on it because of their opposite electrical charges. In consequence one of three things may happen:

1. If the proton does not come too close or remain near for too long a time, the electron may not be disturbed.
2. The electron may be pulled from its normal position in the atom to a higher, ordinarily empty orbit. In other words

the atom is excited. The necessary energy must of course be subtracted from the kinetic energy of the proton which slows down accordingly.
3. The electron may be detached from the atom altogether thereby creating an ion pair. This process takes an even larger quantity of energy from the proton than atomic excitation. See Figure 22.

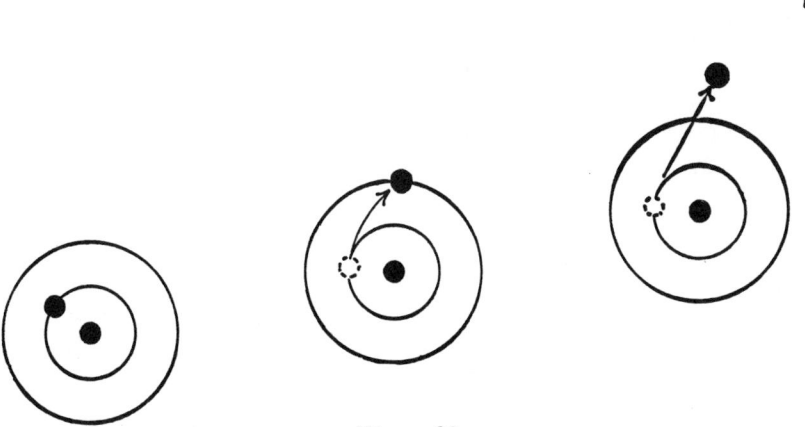

Figure 22.

There is no practicable way to determine the number of excited atoms produced by a passing proton; however, if the irradiated medium is a gas, the ions can often be counted. Such experiments show that in air about one ion pair is produced for every 34 electron volts lost by the proton. Remember, not all of that energy was used to form the ion pair; an unknown fraction of it went to produce excited atoms. It is assumed (although no direct evidence is available) that a similar amount of energy is also required per ion pair in solids and in liquids.

The path of a proton through matter can be made visible. If the material in question is a photographic film, each ion will cause a grain to be activated so that upon development a string of dots marks the passage of the proton.

Charged particle tracks can also be made visible in an instrument called the cloud chamber. This is a vessel containing air saturated with a vapor, usually of alcohol or water. If, immediately after passage of a charged particle, the pressure in the chamber is suddenly lowered, some of the vapor will condense into droplets, using the ions present as nuclei upon which to form. Since these droplets remain distinct for only a few moments, one must photograph them at once. In such photographs pairs of droplets, each corresponding to an ion pair are easily seen and counted.

Since high energy particles may travel for many feet or yards in air at atmospheric pressure, the cloud chamber can be used only for particles whose total energy does not exceed a few MeV. For higher energy particles, one may use a bubble chamber instead. This consists of a vessel filled with a transparent liquid under pressure heated to its boiling point. As soon as a charged particle has passed through the chamber, the pressure is reduced. The liquid starts to boil using the ions as nuclei for bubble formation. These are photographed just as are the droplets in the cloud chamber. Since liquids are about one thousand times denser than air at atmospheric pressure, track lengths are reduced to about a thousandth of those found in the cloud chamber.

There are several significant features of a proton track as seen by any of the methods described. First, it is almost but not quite straight. Each encounter with an electron may deflect the proton slightly from its original direction, but as the mass of the proton is about 1,836 times greater than that of the electron, the change of direction is imperceptable. Nevertheless thousands of such minute deviations may add up to a noticeable waviness of the track. This is most prominent near the end where the proton is going quite slowly. Occasionally, about once in ten thousand times, a proton track contains a sudden sharp angle. The cause of this is a collision with an atomic nucleus.

It is easy to tell which way the proton was going because the ion pairs are closer near the end of the track than near the beginning. Initially the proton is going very fast and its attractive force acts on a given electron for a very short time. In fact, that time may be too short for the electron to move. When the proton has slowed down, the force acts longer and the likelihood of excitation or ionization is correspondingly greater.

When the proton is traveling at nearly the speed of light (a speed which no material object can exceed or even quite equal) the ion density along the track has its lowest possible value. All singly charged particles traveling at that speed, be they electrons, mesons, or protons, all produce the same amount of ionization per unit distance. For that reason such particles cannot be identified from their tracks without additional data.

The track of the deuteron is very similar to that of the proton. Since the deuteron is twice as heavy as the proton, its path is more nearly straight. Also, for a *given initial* energy, the deuteron travels slower; it therefore ionizes more effectively, loses energy faster, and so has a shorter range.

The alpha particle has twice as much charge as the proton and is four times as heavy. Since the ability to excite and to ionize varies as the square of the charge, the alpha particle produces four times as many ions per unit path length as a proton or a deuteron traveling at the same speed. In particular, if an alpha particle is traveling at approximately the speed of light, its ionization density is four times the minimum possible value. Because of its mass and extra charge, the track of an alpha particle, compared to that of a proton of the same initial energy is short and densely ionized.

The most important parameters of a charged-particle track are:

a. range
b. straggling
c. specific ionization
d. stopping power or LET

The range of a particle is the depth to which it will penetrate a material object if it enters along a path perpendicular to the surface. The range depends upon the type of particle, the nature of the stopping medium, and the initial energy. While the mathematical relationship between these quantities is complicated, it can be said that:

a. The greater the energy the longer the range.
b. The heavier the particle and/or the greater its charge, the shorter its range for a given initial energy.

c. The denser the stopping medium, the shorter the range, other things being equal.

The *range* of protons in matter can be determined either mathematically or by experiment. Computation is impractical without a computer; experimental determination, although of limited accuracy, is fairly simple. A possible method consists of placing successive sheets of absorber into a beam of single-energy protons until all are stopped. Figure 23 is a sketch of the apparatus. Tables and graphs of proton range as a function of energy based on both calculation and experiment are now available.

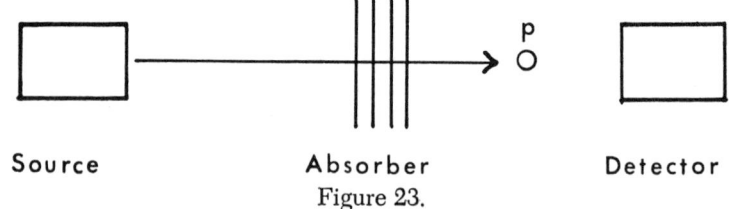

Source   Absorber   Detector

Figure 23.

Suppose that a range measurement is to be made with an apparatus similar to that shown in the figure. Then it will be found that for all thicknesses of absorber up to a certain value, nearly all the protons will penetrate it. If the thickness of the absorber is further increased, the number getting through will fall rapidly to zero. This cutoff is not absolutely sharp however. There is a small range of thicknesses through which the transmission declines. As an analogy, suppose that a hundred cars of the same make and model are each supplied with exactly ten gallons of gasoline and then sent traveling down a certain road. No matter how hard one tries to equalize all the conditions, it will be found that the cars run out of fuel at slightly different spots. In much the same way protons lose their initial supply of energy in paths of slightly different length. This variation of the total distance traveled is called *straggling of the range*.

The cause of straggling is easy to explain. The proton is stopped by thousands of collisions with electrons which take varying amounts of energy. Now it may happen that a particular proton will experience an unusually large number of interactions taking somewhat less than the average amount of energy. If so, that proton will travel farther than usual. Another proton may undergo interactions taking somewhat more than the average amount of energy. It will stop somewhat short.

The number of ion pairs per unit path length is called the *specific ionization* of the particle. If the stopping medium is a gas, the unit of length is ordinarily the centimeter or the millimeter; if a solid or liquid, the micron. Figure 24 shows the varia-

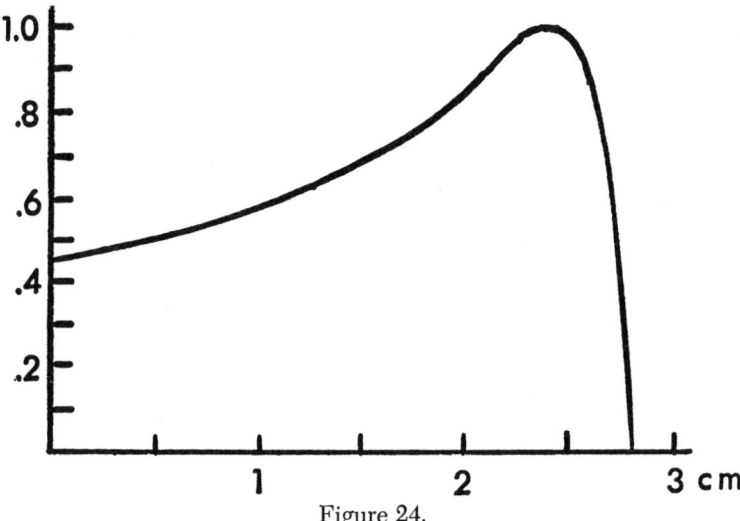

Figure 24.

tion of specific ionization with remaining range; as before mentioned, it becomes greater as the particle slows down until it is nearly stopped. Then of course its ionizing power rapidly falls to zero.

For practical purposes, the terms *stopping power* and *LET* (linear energy transfer) mean the same thing although there is a slight technical difference in their definitions. Both mean energy lost per unit path length. If the stopping medium is a solid or a liquid, the usual unit is the *electron volt per micron.*

Since each ion pair represents an energy loss of a given amount, 34 eV in the case of air, the LET must be directly proportional to the specific ionization. Therefore the curve of one as a function of depth of penetration looks just like a curve of the other except for a scaling factor.

The concept of LET is particularly important in radiation

biology and therapy because it profoundly affects the response of the irradiated tissue. With regard to LET, radiations can be roughly divided into two classes. Electrons and very high speed protons are minimally ionizing and therefore have a low (in fact minimal) LET. Lower energy protons and all multiply charged particles are much more heavily ionizing and hence have a high LET. Medical investigators have tried to take advantage of the change of LET along the path of a proton beam to treat deep-lying cancers without subjecting overlying healthy tissue to excessive dose. So far success has been limited to special situations.

ELECTRONS

The fast electron, whether it be from a decaying nucleus or a high energy accelerator, loses energy in matter just as a proton does—by exciting and by ionizing atoms. However, because it has so little mass, each interaction may change its direction appreciably with the result that its path may be far from straight.

According to the theory of relativity, any moving object is somewhat heavier than an identical one at rest. Its mass m for any velocity v is given by the equation:

$$m = \frac{m_o}{\sqrt{1 - \left(\frac{v}{c}\right)^2}}$$

where m_o is the mass of the object at rest and c is the velocity of light. A more useful formula gives the mass in terms of the kinetic energy T:

$$m = \frac{T}{c^2} + m_o$$

The following table shows the approximate ratio of the masses of moving and stationary electrons for various values of kinetic energy:

kinetic energy	mass/rest mass
0 keV	1
100	1.2
500	2
1 MeV	3
5	10
10	20
50	100

Since at very high energies the moving electron is considerably more massive than the atomic electrons, its track is comparatively straight. As it slows down, its path becomes more and more irregular and often ends in a complete snarl. Because it is so light, the electron moves at nearly the speed of light until its energy has fallen to a very low value. Consequently it is always minimally ionizing except at the end of its journey.

The range of the electron is defined just as for the proton; namely as the thickness of absorber which a beam of particles of given initial energy can penetrate. Several difficulties arise however. For one thing, the range of an electron is not equal to the distance traveled but somewhat less since the electron follows a complicated zig-zag course in passing through the absorber. Furthermore, the straggling is extreme.

Figure 25 shows the fraction of initially monoenergetic electrons that pass through various thicknesses of absorber. The curve can be divided into three distinct regions. In the first region where the thickness of the absorber increases from zero to some small value, the number of exciting electrons increases so that more electrons leave the absorber than enter it. This behavior comes about because some of the electrons released when atoms are ionized have sufficient energy to escape.

In the second region the curve falls from its maximum to a small but nonzero value. This portion of the curve begins and ends horizontally with a long, nearly straight sloping stretch

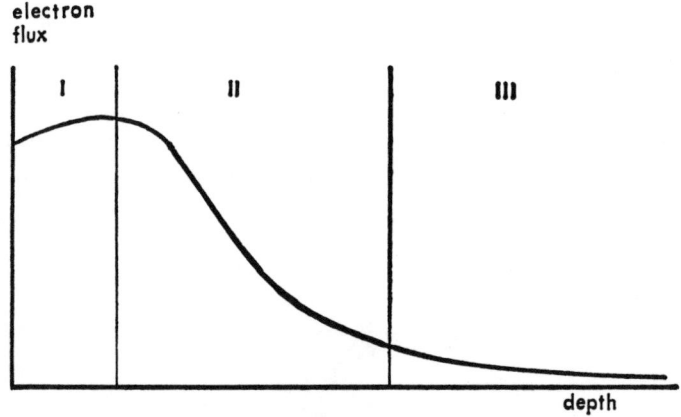

Figure 25.

between. This portion of the curve represents the straggling of the primary beam. Those electrons that are deflected at large angles to the direction of incidence waste most of their energy in going sideways; those that go more or less forward can penetrate comparatively thick absorbers.

The third portion of the curve is a nearly level line that never quite reaches zero. One might at first assume that a few electrons have anomously high penetrating power; that however would contradict physical law as we know it.

What actually happens is that some of the incoming electrons produce x-rays which travel much greater distances in matter than charged particles. Some of these produce photo- or recoil electrons near the back face of the absorber. A number of these escape and enter the detector.

The question is, at what point of the curve can the primary beam be said to be completely stopped? Since regions II and III blend imperceptibly into one another, it is impossible to say. An artificial convention must be adopted. There are several possibilities and in fact two are in common use.

Figure 26 is a repetition of the electron range energy curve with the necessary geometrical additions to determine the *maximum range*, the *extrapolated range*, and the *practical range*.

The line AB is the backward extrapolation of that portion of the curve due to the electron-produced x-rays; the line CD is the

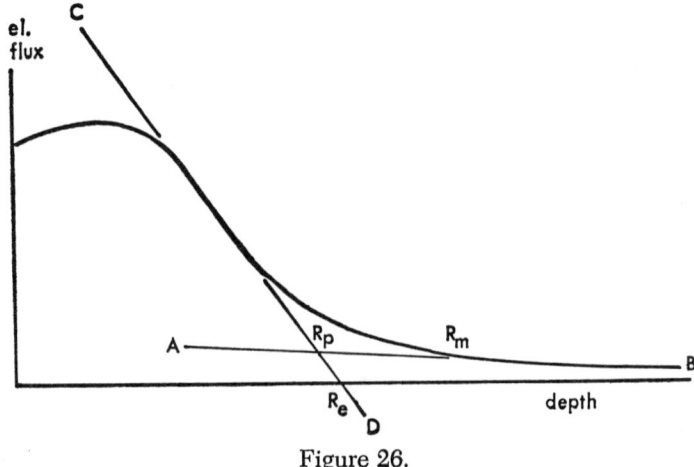

Figure 26.

tangent of maximum slope of the rapidly falling portion of the curve.

The point R_m is where the primary electron beam disappears into the secondary background. The x coordinate of this point is called the *maximum range*. Unfortunately this point can not be located with any precision from most experimental data. Hence while this quantity is of theoretical interest, it is useless in practice.

The point R_e where CD cuts the horizontal axis is much more definite. The x coordinate of this point is called the *extrapolated range*. Unfortunately, the position of this point depends on things other than the initial energy of the electrons. Its use is therefore limited, especially at higher energies.

The point R_p, the intersection of AB and CD, is almost as accurately determined as the point R_e. Its x coordinate is called the *practical range*. This value is a reliable measure of the initial electron energy and can be determined theoretically.

NEUTRONS

The behavior of neutrons in matter is entirely different from that of charged particles. As we have seen, the latter interact almost exclusively with the electrons of the medium; neutrons, which have no electrical charge, are quite unaware of their presence. Hence the neutrons react only with the atomic nuclei. There are several different kinds of interactions that may occur, but from the viewpoint of the radiologist or the health physicist the only two of interest are *neutron capture* and *elastic scattering*.

SCATTERING: If a neutron is moving at high speed, capture is most unlikely. The neutrons usually collide with and bounce off the nuclei of the absorber. For this process there are two distinct kinds of absorber; those that contain hydrogen and those that do not.

If no hydrogen is present, then all the nuclei in the scatterer are several times heavier than the neutrons. These therefore bounce off the former like baseballs off of bowling balls and are scattered in all directions. The nuclei are jostled a bit but do not move any great distance from their initial positions. In each collision the neutron loses a small portion of its energy.

After a few dozen such collisions, the neutrons become *thermalized*, meaning that their kinetic energy has been reduced to that of the surrounding atoms, namely about one fortieth of an electron volt at room temperature. Since the atoms themselves are in motion, thermal neutrons are just as likely to gain energy from a collision as to lose it. Hence thereafter they will ricochet from nucleus to nucleus neither gaining or losing energy. Eventually they will either escape by crossing the boundary of the absorber or be captured by the nuclei within it. If the absorber is a small object, escape is very likely; if it is large and the neutron is originally released in its depths, capture is the more probable fate.

If the absorber contains hydrogen, the situation is considerably changed, the reason being that hydrogen nuclei (which are single protons) have just about the same mass as the neutrons. Since biological material always contains large quantities of hydrogen, only this latter case is of interest to the physician and health physicist.

The collision between the neutron and the proton is comparable to that between two billiard balls. In such an interaction the neutron (which corresponds to the cue ball) loses a substantial fraction of its kinetic energy and in fact may lose all of it. At the same time, the proton shoots off with as much energy as the neutron has lost. The result has been to introduce a fast proton into the absorber. A very few such collisions will thermalize the neutron.

The fast protons produced by neutron irradiation of hydrogenous material differ from those injected by an external beam in two ways. First, they arise at every point within the medium rather than all entering through a surface portal. Hence, by neutron bombardment one can deliver a more or less uniform proton dose throughout the volume of the medium. Second, the protons set in motion by the neutrons have a wide range of energies whereas those from a proton accelerator are usually monoenergetic. The consequence of this fact is that the LET for neutron produced protons is, on the average, uniform throughout the absorbing medium, whereas that of a beam of protons increases with depth. In the first case the biological response is a function of dose only; in the latter, both of dose and position.

NEUTRON CAPTURE: If a neutron remains in the vicinity of an atomic nucleus for any length of time, it is likely to be drawn into it by the nuclear force which (as far as we know) is always attractive. The strength of this force varies enormously for the various nuclei, however. The thermal neutron capture cross section (TNCSS) is a measure of a nucleus' ability of capturing a thermal neutron. This quantity is usually expressed in a unit called the *barn*. Most nuclei have TNCSS's lying between 0.1 and 10 barns although values far beyond these limits are known. He-4 apparently has a TNCSS of exactly zero. C-14 also finds thermal neutrons distasteful as evidenced by a TNCSS of about a millionth of a barn.

At the opposite extreme are several nuclides with enormous cross sections, a few of which are:

Au-198	26,000 barn
Cd-113	27,000
Sm-149	41,500
Gd-157	240,000

Of these, cadmium is used in nuclear reactors to keep the quantity of thermal neutrons under control. If the number gets too large, cadmium rods are thrust inward to absorb the excess; if the number is too low, the control rods are partly withdrawn. Although cadmium has by no means the largest known TNCSS, it is comparatively cheap and has suitable physical and chemical properties. Materials with higher cross sections are either very rare or have other disqualifying characteristics.

When a nucleus captures a neutron it becomes a different isotope of the same element. C-12 becomes C-13, Co-59 becomes Co-60 and so on. In each case Z remains the same while N increases by one. The new nucleus is always in a very highly excited state since the lowest unfilled level in any nucleus has a binding energy of six to twelve million electron volts.

From the standpoint of radiation safety the neutron capture

process is of little concern; that is to say, it would be most unlikely for a person to find himself in a region where the thermal neutrons are the principal hazard. Fast neutrons, if present, and gamma rays would be a far greater risk.

At least one attempt was made to use thermal neutrons therapeutically. Victims of certain types of brain cancer were given massive doses of boron which would collect in the tumor but not in the healthy tissue of the brain. The patient's head was then placed in a beam of thermal neutrons. Boron, which has a high TNCSS, would then undergo the reaction:

$$^{11}_{5}B_6 + ^{1}_{0}n_1 \rightarrow ^{12}_{5}B^*_7 \rightarrow ^{12}_{6}C^*_6 \rightarrow 3\alpha$$

Because of their high LET the alphas were expected to have a devastating effect on the tumor.

This mode of treatment, so beautiful on paper, didn't work too well in practice. Boron in sufficiently large quantities proved somewhat toxic to the patient; the tumors did not take it up as well as one hoped; and the amount of boron that found its way into healthy tissue was excessive. While the outlook is not too promising, it may be that future investigations along similar lines may be more satisfactory.

PHOTONS

What is a photon? About the best we can say is that a photon is one of the fundamental units of the universe, an *object* uniquely different from all others. As is true of all basic entities, it can be defined only by describing its *appearance* and behavior.

Photons are always in motion. In empty space they move in straight lines with a speed of 2.997×10^8 meters per second. When travelling in matter they may move a bit more slowly and their trajectories may be curved. A sudden change of direction, known as reflection, may occur at boundary surfaces. In any case a stationary photon is impossible. Hence the sometimes seen statement, "The rest mass of a photon is zero", is a trifle silly.

Unlike material particles, photons are easily created and destroyed. A lamp within a room produces countless billions of photons every second nearly all of which are absorbed moments later by the walls or other objects within. The sole evidence of the photons' ever having existed is a slight warming of the absorber. Only those that escape into the vacuum of outer space will last any length of time.

By comparison, material particles are much more stable. In fact it was once thought that they could be neither created nor destroyed. We now know better, but special effort is required, and certain types (i.e. electrons, protons, and atomic nuclei) are likely to last indefinitely.

Every photon contains energy. It is sometimes said that a photon *is* energy traveling through space. Be that as it may, whenever a photon is produced, a certain amount of energy is abstracted from the generating system. When the photon is absorbed, the same amount of energy reappears in another form, often (but not always) as heat.

There is no known limit to the amount of energy that a photon may have; photons of all energies from just above zero to several billion electron volts have been observed and even produced by man. Presumably a single photon could have as much energy as Grand Coulee Dam.

Material objects have definite size and shape; one can say with considerable exactitude that a proton is a sphere with such and such a radius. No comparable statement about photons is possible. However in some quasi-literal fashion the concept of size must be associated with them. Radio waves are as big as a barn or even bigger; x-ray photons can slip between the atoms of every substance.

A measure of photonic size is its wavelength. The word *wavelength* is used because under certain conditions, photons create patterns of ripples much like the waves on a pond. It should be remembered however, that while photons *may* produce wave patterns, they do not *have* to. In particular, photons in the x- and gamma ray regions almost never exhibit wave properties in practical applications although such can be demonstrated in special laboratory situations.

The wavelength of a photon and its energy are inversely

proportional. If the energy E is expressed in keV and the wavelength λ in angstroms ($1 A = 10^{-10}$ m) then:

$$\lambda = \frac{12.4}{E}$$

While the properties of photons change with energy, there are no sharp, sudden transitions in their properties. If we compare two photons whose energies differ by, say, ten per cent, they will be found to be very much alike. However if we compare two that differ in energy by 1,000 per cent, they will be quite different. Because of this gradual transition in properties, it is impossible to divide the photon spectrum into sharply distinct regions. All such attempts are arbitrary, and are done chiefly as a matter of convenience.

Not only does the photon lack definite size and shape; it has no definite position either except at the moments of creation and annihilation. In fact certain experiments can be explained only by assuming that the photon, while in flight, was in at least two different places at the same time.

Every photon contains—or consists of (?)—electric and magnetic fields. Because of that fact, photons are often referred to as *electromagnetic radiation*. They exert a force on any object that contains electric charge. Even an object, such as a neutron, which as a whole is electrically neutral, may interact with a photon via the magnetic field produced by circulating charges.

THE ELECTROMAGNETIC SPECTRUM: Since photons have such a wide gamut of properties and have so many important effects, it is useful to classify them, artificial as the boundary points may seem.

The first, simplest, and most obvious classification is with respect to human vision. The normal eye responds to photons with wavelengths between 4,000 and 7,000 angstroms or (what is saying the same thing) energies between 1.7 and 3.0 eV. Photons in this range constitute what the layman calls *light*. Photons of lower energy and longer wavelength are called *infrared* while those on the other side of the visible portion of the spectrum are called *ultraviolet*.

The Interaction of Radiation with Matter

Logically the terms *infrared* and *ultraviolet* should be applied to all non-vision-inducing photons; as a matter of convenience these words are confined to those comparatively near the visible region.

Photons may also be classified as *ionizing* or *non-ionizing*. Because of their electromagnetic fields, photons exert force on every electron they pass near as already mentioned. If the photon has sufficient energy, it may pull an electron out of an atom thereby ionizing it. Since the binding energy of the valence electrons (which varies from substance to substance) usually lies near or above five electron volts, infrared and visible photons can not produce ions whereas nearly all ultraviolet do.

The effect of radiation on matter varies profoundly according as the photons are or are not ionizing. In particular, biological tissue can stand heavy exposures of non-ionizing radiation—consider for example the heat and light from the sun. Small quantities of ionizing radiation on the other hand are extremely dangerous; the relatively small amounts of solar ultraviolet light which reaches the surface of the earth can cause severe skin reactions.

The electromagnetic spectrum may also be divided according to the mode of generation. Photons with wavelengths of several centimeters or more can be produced by large, man-made devices such as vacuum tubes, magnetrons, transistors and the like; photons with wavelengths much less than a millimeter are produced only by molecules, atoms, or subatomic particles. Those in between may be produced either way.

Those photons which can be produced by macroscopic generators are much used for television, radar, and the like. Hence, streams of such photons are usually referred to as *radio waves;* the term *infrared* is restricted to the region between the radiowaves and the visible spectrum. Sometimes the word *microwave* is used to distinguish an intermediate zone between radio and infrared radiations. Obviously the boundary between the radio and the infrared regions shifts to a shorter wavelength value with every advance in technology.

Still another division of the electromagnetic spectrum results from the fact that the atoms of all solids and liquids are about one angstrom apart. One would surmise therefore that photons which are *bigger* than one angstrom would be unable to pass through matter while those smaller would not be seriously hindered. In other words, a piece of matter would be a photon sieve. This sort of thing in fact happens. However there are special circumstances whereby very large photons may pass through certain media. For example, visible light passes easily through glass, water, and a number of other substances. The reasons are well understood but too complex to explain here. Transparency is a limited phenomenon however. It is usually true that photons with wavelengths greater than one angstrom can not penetrate solids and liquids. Comparatively few materials are exceptional and these for only limited ranges of wavelengths.

Photons with wavelengths much less than one angstrom are called *x-rays*, and can penetrate every known substance, being stopped only if they collide with an electron or an atomic nucleus.

Between the far ultraviolet and the x-ray region are those photons with wavelengths about equal to one angstrom. These pass through matter only with difficulty. They are called *Grenz rays*, *Grenz* being the German word for *boundary*. This region is significant because all matter is densely opaque to far ultraviolet light. The special conditions which can sometimes produce transparency for visible and infrared rays never exist in the far ultraviolet region. There, even gases are highly absorptive, so that experiments must be conducted in a vacuum. The Grenz region is that wherein total opacity gives way to universal transparency.

Because all matter is opaque to ultraviolet light (only that just beyond the visible spectrum being an exception) such radiation is not usually hazardous although it is ionizing. Only those ultraviolet photons just above the visible spectrum can penetrate the atmosphere or the dead outer layers of the skin. These few however can produce severe burns as any outdoorsman knows, especially if he is blonde or redhead.

A reasonable—but arbitrary—division of the electromagnetic spectrum is:

Kind of Radiation	Wavelength	
Radio waves	infinite to 10 cm	
Microwaves	10 to 0.1 cm	non-ionizing
Infra red	0.1 cm to 7,000 A	
Visible radiation	7,000 to 4,000 A	
Ultraviolet radiation	4,000 A to 10 A	
Grenz radiation	10 A to 1.0 A	ionizing
X radiation	less than 1.0 A	

The distinction between x-rays and gamma rays is somewhat awkward. According to current usage:

A photon is an x-ray if and only if (a) its wave length is short enough to enable it to penetrate all matter to some extent and (b) it is not produced by an atomic nucleus or by a fundamental particle reaction.

A gamma ray is a photon produced by an atomic nucleus or a fundamental particle reaction.

X-rays are ordinarily produced in either of two ways; by decelerating a fast electron near an atomic nucleus or by certain electronic rearrangements within an atom. Those of the first type constitute what is known as *Bremsstrahlung* (German for *braking radiation*); those of the latter type are called *characteristic radiation*.

As far as the definition goes, a gamma ray could occur anywhere in the electromagnetic spectrum. Although the thought of reading a newspaper illuminated by visible gamma rays is a bit startling, it is permissible within the scope of the definition.

It is worth remembering that x-rays and gamma rays of the same energy are identical in all their properties and are therefore indistinguishable. The difference in name pertains only to origin. There is some tendency to distinguish x- and gamma rays according to energy, especially in the foreign literature. According to this informal convention, a photon with energy less than about one MeV would be an x-ray and one with higher energy a gamma ray. Thus, it is often said that a betatron or a linear accelerator emits gamma rays of, say 24 MeV, whereas in strict accuracy the photons from such a machine should be called x-rays. In any case the distinction is rarely important except in classrooms or certifying examinations.

Photons of all energies may interact with matter. In the visible range, for example, the usual interactions are absorption, transmission, reflection, and refraction. In this book however we are concerned only with those in the x-ray region. They undergo several important types of reactions including:

1. Photoelectric absorption
2. Compton scattering
3. Pair and triplet production
4. Nuclear photodisintegration

PHOTOELECTRIC ABSORPTION: In this process a photon which comes near a bound electron disappears completely with ejection of the electron from the atom. The initial energy of the photon is divided into two portions: The first is used to free the electron from the atom and is obviously equal to the binding energy of the electron in question. The remainder is transferred to the electron as kinetic energy. That is:

$$E = E_{binding} + E_{kinetic}$$

For example, suppose that a 100 keV photon is absorbed by a K electron in lead whose binding energy is 88 keV. Then 88 keV is expended in removing the electron, and the remaining 12 keV is used to give it a departing speed of 6.39×10^7 m/sec or 21.3% of the velocity of light.

Photoelectric absorption is most probable for photons of low energy in substances of high atomic number. Thus if photons of 20 keV strike a lead absorber, 99% of them will undergo photoelectric absorption in less than half a millimeter, whereas if a block of carbon is irradiated with 1 MeV photons, the number of such events will be almost nil.

The variation of the photoelectric effect with Z is crucially important in diagnostic radiology. Usually one uses photons of such energy that this process almost never occurs in soft tissue but is still quite frequent for bone with its calcium and phosphorus content. The result is that soft tissue is much more transparent than bone.

Many body cavities can be studied by filling them with a material containing a high-Z element such as iodine or barium. In this way the bladder, GI tract, cerebral ventricles, and the blood vessels can be rendered visible.

COMPTON SCATTERING: Compton scattering differs from photoelectric absorption in that the photon does not disappear. Instead it ricochets off the electron concerned. This reaction is comparable to the collision of two spheres of different mass. In this process the photon behaves very much like a material particle. It strikes the electron which is at rest (or comparatively so) and both objects, the electron and the photon, recoil in different directions.

The uniquely different property of this reaction is the way in which the photon loses energy. If it were a material particle, it would slow down; that being impossible, its wavelength increases.

The strength of the Compton interaction depends but slightly upon the atomic number of the material irradiated. It is instead proportional to the number of electrons per gram, a quantity which is about the same for all substances except hydrogen.

The intensity of the Compton effect falls off as the energy increases but much more slowly than that of the photoelectric absorption. Hence photoelectric absorption is overwhelmingly predominant for very low-energy photons but insignificant compared to the Compton effect when the photon energy becomes sufficiently high.

The cross-over energy, namely that where photoelectric absorption and Compton scattering are equally strong, depends upon the atomic number of the material. For water—a typical low-Z material—it occurs at 26 keV. For lead it is much higher, at about 500 keV.

Another important difference between the photoelectric and the Compton effects is that the former chiefly affects those electrons whose binding energies are not too much less than the photon energy, the latter involves electrons with binding energies very much less than the energy of the photon. Many authors say that the Compton effect takes place between photons and *free* electrons, but such statements are not quite correct.

PAIR PRODUCTION: When a photon passes near an atomic nucleus, it may be transformed into an electron and a positron. In this reaction the photon disappears and both particles come into being at a point near the nucleus. 1.02 MeV of the photon's energy is used to make up the rest mass of the two particles; the remainder is distributed in any proportion between the two as kinetic energy. Obviously (since energy must be conserved) only those photons having at least 1.02 MeV of energy can produce electron-positron pairs.

If the photon has an energy greater than 2.04 MeV, pair production may also occur when a photon passes near an electron. In this case the electron which initiated the process also recoils at high energy so that the photon has apparently produced three particles. Consequently this reaction is called—rather inaccurately—triplet production.

As mentioned, the probability of pair production is zero for photons with energy less than 1.02 MeV. Thereafter the probability of this process increases rapidly with both photon energy and atomic number of the absorber.

PHOTONUCLEAR DISINTEGRATION: Just as a photon may remove an electron from an atom, so it can knock a proton or a neutron from an atomic nucleus if it has sufficient energy. For most nuclides, at least eight MeV are required. Since the residual nucleus is often radioactive, areas and objects exposed to high-energy photons may be dangerous for a period of time. Fortunately most of the radionuclides produced are very short lived. The neutrons released may also be a hazard. Hence whenever dealing with photons of more than ten MeV, one should measure the quantity of neutrons and radioactivity produced before permitting human exposure.

Two typical reactions are:

$$^{16}O + \gamma \rightarrow {}^{15}O + n$$
$$^{14}N + \gamma \rightarrow {}^{13}N + n$$

In the first, the threshold energy is 15.6 MeV, and the half life of O-15 is about two minutes. In the second, the threshold energy is 10.5 MeV and the half life of N-13 is about ten minutes. Hence the atmosphere itself may become radioactive when exposed to x rays of sufficiently high energy.

Chapter XIV DOSE AND EXPOSURE

IN THE FIELD OF radiation physics, the word *dose* means: *The amount of energy absorbed per gram of the irradiated object from the incident radiation.* The most natural units of dose would be either the erg/gram or the joule/kilogram. As neither of these is of convenient size for the physician or the biologist, a special unit, called the *rad*, was created. It is equal to 100 erg/gram. One advantage of this unit is that under most clinical circumstances, an exposure of one roentgen (a unit long in use which will be discussed later) gives a dose of about one rad.

The dose delivered to any specimen may be determined either by measurement or by calculation. In most cases measurement is impractical and calculation tedious.

It rarely happens that the dose is uniform throughout the irradiated object; usually it is different at every point—which immediately raises the question: What is meant by "dose at a point?" A point being infinitely small, the amount of matter within it is zero, and the energy absorbed there is also zero regardless of the total amount of radiation incident upon the body. The dose is then: zero/zero, a meaningless concept.

To get around this barrier, the notion of dose *at a point* is revised to mean the dose to a small quantity of matter surrounding the point in question. This test sample is chosen as small as possible and is usually thought of as a sphere centered on the point. Two conditions are imposed on the size of the test sample: it must be so small that the dose is practically uniform across it, and yet big enough for accurate, reliable measurements. Usually these two conditions allow a range of usable sizes; when not, no dose measurement is possible.

There is only one way to make a truly direct measurement of dose. In nearly all substances the absorbed energy appears almost entirely as heat since the energy used for other physical and chemical changes is a tiny fraction of the total. Hence dose measurement is reduced to the measurement of heat absorption, meaning in practice a measurement of temperature variation.

The rise of temperature associated with an increase of heat content is given by the equation:

$$\Delta t = \frac{\Delta H}{ms}$$

where Δt is the change in temperature, ΔH is the heat added, m is the mass of the body, and s is its specific heat. The usual units are degrees Celsius, calories, and grams.

How large a temperature change will a dose of one rad induce in biological material? Since an erg is equal to 2.389×10^{-8} calorie, a rad equals 2.389×10^{-6} calorie/gram. As biological tissue is mostly water, its specific heat is about 1.0. Therefore:

$$\Delta t = \frac{2.389 \times 10^{-6}}{1 \times 1} = 2.389 \times 10^{-6} \, °C$$

Such small temperature changes are just about impossible to measure. Even for doses a thousand times greater the technical difficulties are considerable. Nevertheless appropriate apparatus has been constructed in a number of laboratories.

A second type of dose measurement is based upon the fact that many chemical reactions are induced by radiation with yields accurately proportional to the total dose. Two difficulties arise however. The yield of reaction products is always small, so that chemical dosimeters work only for high doses. Second, every conceivable reaction that might be used presents a host of unique difficulties. Some lead to products that cannot be measured accurately in the quantities available; some require impossible standards of cleanliness. Others are erratic in spite of all reasonable precautions. Only the *Fricke Dosimeter* has found wide acceptance. It consists of a solution of ferrous sulfate which, upon irradiation, becomes ferric sulfate. The tiny quantity of the latter can be determined fairly conveniently with a spectrophotometer.

Using the Fricke Dosimeter is not easy; only large doses can be measured; extreme but practicable chemical purity is essential. These conditions observed however, reproducible data are obtainable. In many laboratories the Fricke Dosimeter is used to calibrate the secondary instruments used for day-to-day dose determinations.

EXPOSURE

For routine determinations of dose when dealing with x- and gamma rays, the most common procedure is to measure another quantity called *exposure* and then calculate the dose from that. The usual unit of exposure is the *roentgen* which is defined by the following procedure:

Let a quantity of air surround the point at which the exposure is to be determined. This volume of air must be divided into two parts:

a. a small volume dV centered at the point P upon which the x or gamma radiation is permitted to act,
b. a larger volume of air V surrounding the above.

The outer quantity of air must not be irradiated and must be thick enough to stop all electrons ejected from the inner volume. See Figure 27. No material barrier of any sort may be placed between the two quantities of air; the surface of separation must be *mathematical*—not physical—in nature.

The radiation acting on the inner volume of air will release fast electrons partly by the photoelectric effect and partly by Compton scattering. If there are any photons with energies greater than 1.02 MeV, electrons and positrons may also be produced by pair production. These electrons will move through both volumes of air producing ions as they go until they are brought to rest. Let all the ions of one sign (either positive or negative as convenient) be collected. Ions are collected from both air volumes. If the total charge so obtained amounts to 2.58×10^{-4} C/kg of air irradiated, then the exposure is—by definition—one *roentgen*.

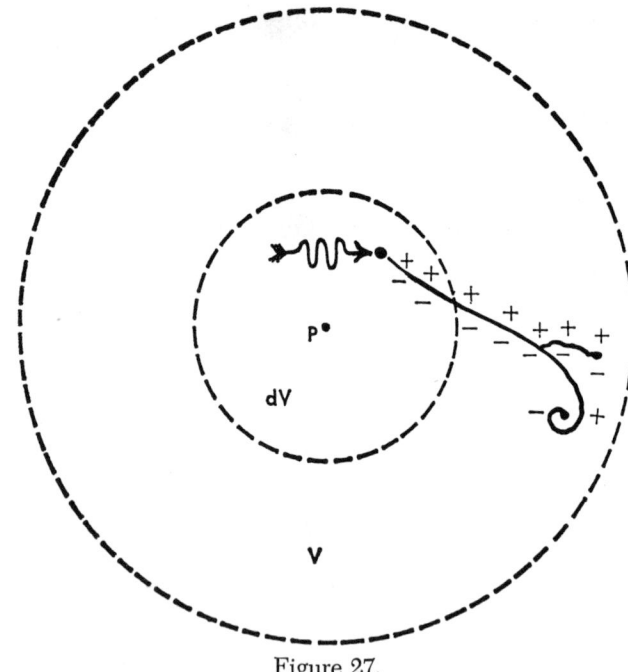

Figure 27.

The experimental setup above described is impossible to achieve in practice. However it can be approximated by an instrument called the *free air ionization chamber*. Such chambers are big, expensive, and difficult to operate; hence they are usually found only in the standard laboratories of various nations and in the calibration laboratories of large institutions. They are used chiefly to calibrate more convenient instruments for routine measurements.

Day-to-day exposure measurements are frequently made with a *thimble chamber*. See Figure 28. It consists of five essential parts:

1. A thimble-shaped cap of *air-equivalent* material with an electrically conducting inner lining which serves as the outer electrode.
2. An inner electrode, also of air equivalent material. (Truly air equivalent material with suitable physical, chemical, and electrical properties does not exist. In practice therefore, the outer cap is usually made of a plastic and the

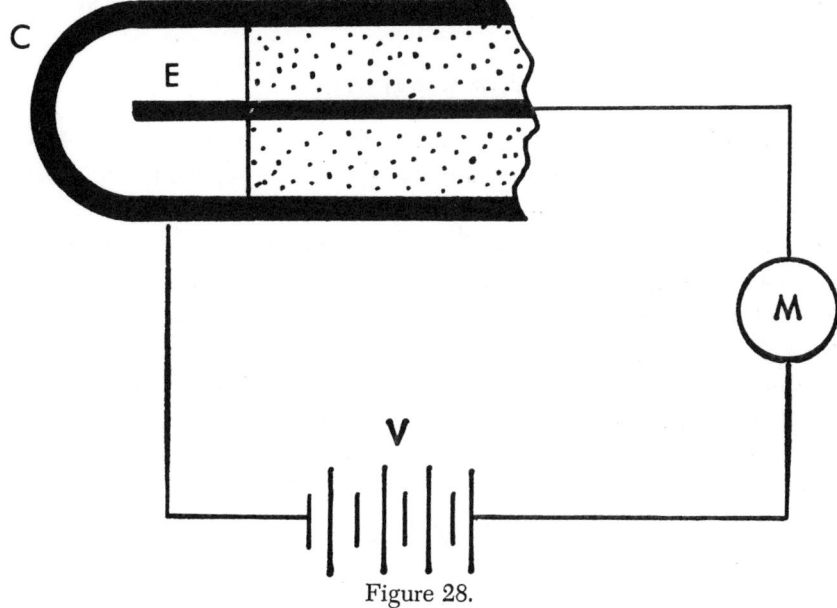

Figure 28.

inner electrode of aluminum. The relative quantities of each substance are so chosen that the combination is very nearly air equivalent over the range of x-ray energies for which the instrument is designed.)
3. A volume of air between the two electrodes.
4. A source of voltage such as a battery, power supply, or capacitor.
5. A read-out instrument, either an electrometer or a pico-ammeter.

In operation photons eject fast electrons from the atoms of the electrodes. A few are also released in the air volume itself, of course, but these are a negligible fraction of the total. Of the fast electrons produced, a certain fraction passes through the inner air volume ionizing as they go. The ions thus formed are collected and sent to the measuring instrument.

If the read-out instrument registers the total charge released by the radiation, its reading is proportional to the total exposure. If, on the other hand, it measures current, its reading is proportional to the exposure rate. On commercially available instruments, the device is usually scaled directly in R or mR for total exposure. Rate meters are scaled in R/min, mR/hr etc. as convenient.

THE F FACTOR

Exposure measurements can also be made in liquid and solid bodies provided it is possible to place an exposure meter of some sort at the point in question. Liquids seldom present a problem, it sufficing to surround the instrument with a fluid-proof cover. In solids, a suitable cavity must be available. When direct measurement is out of the question, one may determine the exposure at the surface of the body and then calculate the value desired using *depth-exposure* tables, or (as they are usually called) *depth-dose* tables.

When dealing with material objects, one is rarely interested in the exposure but rather in the dose. This latter datum can be obtained by multiplying the exposure by what is known as the *f factor*:

(Dose in rads) = (Exposure in Roentgens) × (f factor)

For soft tissue, the value of f lies between 0.9 and 1.0 for all qualities of x radiation; for bone, the f factor is almost 4.5 for low energy x rays and falls to less than 0.95 for x and gamma photons in the supervoltage region. Tables of f factors can be found in various publications including NBS Handbook #85. The variation of f with energy is shown in Figure 29.

DOSE FROM INGESTED CHARGED PARTICLE EMITTERS

The dose measurement procedures previously described are most suitable for external beams of radiation. When unsealed radioactive materials are introduced into the body so that they combine intimately with body fluids and diffuse throughout the tissues, other methods must be used. Sometimes it is possible to make measurements either by inserting appropriate instruments into the body or by taking samples of tissue or fluids. Often, one must rely upon calculations. These are fairly easy for alpha and beta emitters.

We will illustrate the method by a number of special examples. The results can be reasonably accurate if one knows:

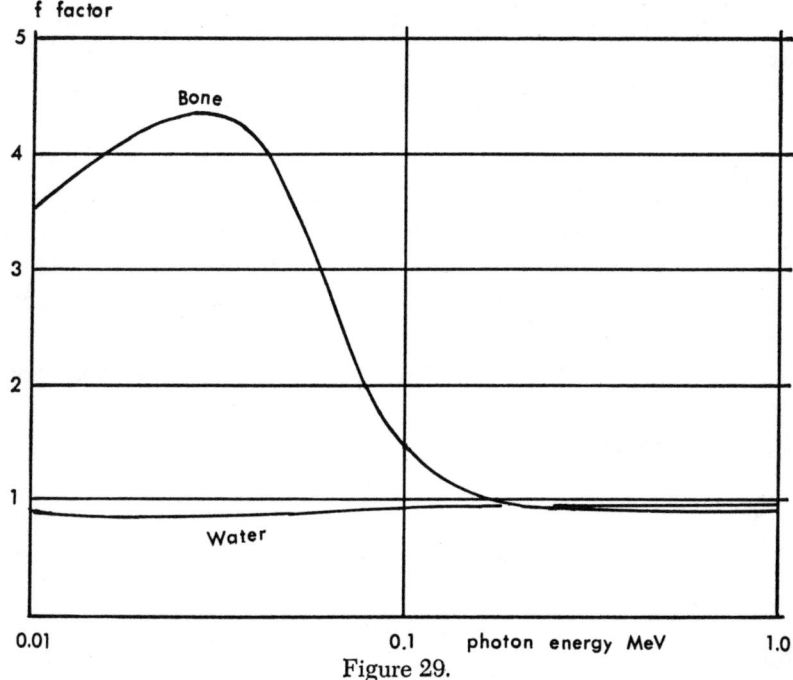

Figure 29.

1. The physical and chemical properties of the radioactive substances used.
2. The quantity of radioactivity in the body.
3. The mass of tissue involved.
4. The distribution of the radionuclide.
5. That the dimensions of the volume impregnated with the radionuclide are considerably larger than the range of the emitted particles. This last condition is almost always fulfilled in larger animals and man. In laboratory animals such as mice, it may well not be.

Unfortunately in many practical cases the required information may not be known at all accurately. When such is true, the results are correspondingly in doubt.

EXAMPLE 1: A certain man drinks one millicurie of tritiated water. (I know, that is a silly thing to do, but we need a simple problem to begin with). If he weighs 70 kg, and the material is excreted exponentially with a half life of ten days, find the initial dose rate and the total dose.

In this problem the requirements for an accurate calculation are met. In particular, one may plausibly assume that the tritium is uniformly distributed since it is mingled with the body water. From standard tables we learn that the physical half life of tritium is 12.3 years and the average energy of the beta particle is 5.5 keV.

This problem can be solved at once with formulas given in many text books; unfortunately, while convenient, these give no insight into the underlying physical principles. Let us instead go about it logically:

First, we must determine the initial dose rate. If the body contains one millicurie of any radionuclide, then by definition 3.7×10^7 nuclei disintegrate every second. In the process each gives off a beta particle.

Since on the average each ejected electron has a kinetic energy of 5.5 keV, the total energy release per second is:

$$3.7 \times 10^7 \frac{\text{dis}}{\text{sec}} \times 5.5 \frac{\text{keV}}{\text{dis}} = 2.035 \times 10^8 \frac{\text{keV}}{\text{sec}}$$

But since:

$$1 \text{ keV} = 1.602 \times 10^{-9} \text{ erg}$$

the total energy released per second is:

$$2.035 \times 10^8 \frac{\text{keV}}{\text{sec}} \times 1.602 \times 10^{-9} \frac{\text{erg}}{\text{keV}} = 0.3260 \frac{\text{erg}}{\text{sec}}$$

It is much more convenient to deal with a longer period of time such as the day:

$$0.3260 \frac{\text{erg}}{\text{sec}} \times 86,400 \frac{\text{sec}}{\text{day}} = 2.817 \times 10^4 \frac{\text{erg}}{\text{day}}$$

The energy to one gram of tissue per day is:

$$\left(2.817 \times 10^4 \frac{\text{erg}}{\text{day}}\right) \div (70,000\text{g}) = 0.4024 \frac{\text{erg}}{\text{g-day}}$$

Since one rad equals 100 ergs per gram:

$$\text{dose} = 0.004024 \frac{\text{rad}}{\text{day}} = 4.024 \frac{\text{millirad}}{\text{day}}$$

A word as to accuracy: the calculations above were carried to four significant figures since that is perfectly easy to do with a desk calculator or a circular slide rule. Apart from arithmetic blunders, this procedure assures that the error arising from successive rounding off will be less than one tenth of one per cent. However, the final answer is no better than the least accurate of the input data. If the value for the amount of ingested tritium is off by, say seven per cent, so is the final answer. In problems of this sort, the accuracy of the input data is often no better than plus or minus ten per cent and may be worse. In the present example, the best that can be said with confidence is that the daily dose is about four millirad.

Total dose can be calculated only if one knows how the quantity of radioactive material in the body varies with time. For this problem we have assumed exponential elimination. That means that the rate of excretion is proportional to the total body burden. It also means that in a given period of time (called the biological half life) fifty percent of the amount initially present will be excreted. During each successive half life, half that remaining from the previous period will disappear and so on. Mathematically, exponential excretion follows the same pattern as radioactive decay.

We make the assumption of exponential excretion partly because many substances are in fact so eliminated and partly because it simplifies the calculation.

Since the physical half life of tritium is over twelve years and (in the present case) the biological half life is only ten days, only the latter datum is relevant for dose calculation.

It can be proved that in the case of exponential loss (whether by physical decay or biological excretion or both) the total dose is equal to:

Dose = 1.44 × (initial dose rate) × (effective half life)

Whence in the problem above:

Dose = 1.44 × 4.042 × 10 = 58.2 millirad

In view of the probable uncertainties in the input data, one can say only that the total dose is about sixty millirad.

Example 2: What is the initial beta dose rate and the total beta dose to the thyroid gland of a patient who has been given ten microcuries of iodine-131? The mass of the gland is

40 g, the biological half life has been determined to be 20 days for this particular patient, and the uptake is 30 per cent.

Unlike the preceding example, which is of little practical importance, this one is typical of the kind occurring in nuclear medicine. Unfortunately the input data are subject to serious errors, namely:

1. It is impossible to get a reliable estimate of the mass of the thyroid gland without removing it surgically and weighing it. Even experienced physicians may miss by fifty per cent or more when trying to estimate the size of the gland in a patient.

2. It can never be taken for granted that iodine-131 is uniformly distributed throughout the thyroid tissue. The reverse is more likely to be the case. If so, the dose values calculated are only averages. At any given point the dose may be either higher or lower than this value.

3. The uptake, while measurable, is hard to determine accurately. Needless to say, the final answer can be no better than this datum.

4. Excretion may follow a somewhat irregular pattern. The fact may be determined by daily measurements on the patient, but such a procedure takes more time and effort than is ordinarily justified.

5. In addition to the beta dose, I-131 also emits gamma rays. Calculation of their effect is exceedingly difficult, and the results are dubious. Experience shows that the gamma dose to the gland is roughly equal to ten per cent of the beta dose. However, while the beta dose is pretty much limited to the gland itself, the gamma rays penetrate to other parts of the body.

Assuming, as we must, that our input data are correct, we solve the problem as follows:

From a suitable reference we find that the average energy of the beta particle given off by I-131 is 188 keV and that this nuclide has a physical half life of 8.07 days.

The total number of disintegrations per day in the gland is:

$$10 \mu\text{Ci} \times 0.3 \times 3.7 \times 10^4 \, \frac{\text{dis}}{\mu\text{Ci-sec}} \times 86{,}400 \, \frac{\text{sec}}{\text{day}} = 9.590 \times 10^9 \, \frac{\text{dis}}{\text{day}}.$$

The total energy released per day is:

$$9.590 \times 10^9 \frac{\text{dis}}{\text{day}} \times 188 \frac{\text{keV}}{\text{dis}} \times 1.602 \times 10^{-9} \frac{\text{erg}}{\text{keV}} = 2{,}888 \frac{\text{erg}}{\text{day}},$$

which converted into rad/day is:

$$\frac{2{,}888 \frac{\text{erg}}{\text{day}}}{40\text{g} \times 100 \frac{\text{erg}}{\text{g-rad}}} = 0.722 \frac{\text{rad}}{\text{day}}.$$

In order to determine the total dose, one must know how the quantity of nuclide present changes with time. Two processes contribute to removing the I-131 from the gland: radioactive decay and biological excretion. In the present problem the half lives of these two processes are not too unequal, being 8.06 and 20 days respectively. Hence both must be taken into account.

There exists an effective half life which can be calculated from the other two by means of the formula:

$$T_{1/2}{}^{\text{eff}} = \frac{T_{1/2}{}^{\text{phy}} \times T_{1/2}{}^{\text{bio}}}{T_{1/2}{}^{\text{phy}} + T_{1/2}{}^{\text{bio}}}$$

Whence:

$$T_{1/2}{}^{\text{eff}} = \frac{8.07 \times 20}{8.07 + 20} = 5.750 \text{ day}$$

Using this value, we get a total dose of:

$$\text{Dose} = 1.44 \times 0.722 \frac{\text{rad}}{\text{day}} \times 5.750 = 5.978 \text{ rad}$$

As stated before, while we carried four significant figures for computational accuracy, the actual value may deviate greatly from the one calculated because of errors in the input data. In a real case we could say on the basis of the above calculation that the total dose is in the vicinity of six rad.

EXAMPLE 3: A woman weighing 60 kg swallows two microcuries of radium-226, ten per cent of which becomes deposited in the skeleton. What is the yearly dose to the bone?

To solve this problem several assumptions must be made:

1. Although there is considerable uncertainty both in the value of the quantity ingested and in the uptake, we must accept the figures given which in real life would be determined by making a variety of measurements and estimates.

2. It must be assumed (unless there is experimental evidence to the contrary) that the radium is uniformly distributed throughout the skeleton. If this condition is false (as it probably is), then the calculated dose will be only an average value.

3. Some assumption must be made concerning the radioactive decay products of radium. In the absence of information to the contrary, one may assume that they are removed from the bone before contributing significantly to the dose.

4. The mass of the skeleton can only be estimated. For people of average build, it is roughly ten per cent of the total body weight.

5. Radium-226 is a pure alpha emitter. 95% of the time the outgoing particles have an energy of 4.79 MeV, the remainder of the time, 4.61 MeV. The average alpha energy is therefore 4.78 MeV.

6. The half life of radium-226 is 1602 years so that during the life time of the patient its activity remains unchanged. The biological half life of radium is also long. Not having been told otherwise we shall assume that the patient's body burden remains constant for the rest of her life.

On the basis of the above assumptions we calculate as follows:

$$2 \; \mu Ci \times 0.1 \times 3.7 \times 10^4 \; \frac{dis}{\mu Ci\text{-sec}} \times 86400 \; \frac{sec}{day} \times 365.2 \; \frac{day}{yr}$$

$$= 2.335 \times 10^{11} \; \frac{dis}{yr}$$

The total energy deposited is:

$$2.335 \times 10^{11} \; \frac{dis}{yr} \times 4.78 \; \frac{MeV}{dis} \times 1.602 \times 10^{-6} \; \frac{erg}{MeV}$$

$$= 1.788 \times 10^6 \; \frac{erg}{yr}$$

Whence the dose rate is:

$$\frac{1.788 \times 10^6 \; \frac{erg}{yr}}{6000 \; g \times 100 \; \frac{erg}{g\text{-rad}}} = 2.98 \; \frac{rad}{year}$$

In view of the uncertainties mentioned, all that can be said is that the patient is receiving an annual dose in the neighborhood of three rad.

INDEX

A

Absorption, photoelectric, 128–129
Activity, 76–78
Alkali metals, 53–54
Alpha emission, 10, 67, 83–86
Alpha emitters, 98–100
Alpha particle, 67, 84, 110, 113, 122
Alpha track, 113
Aluminum-27, 102
Amorphous solids, 43
Ampere, 8–9, 23
Ampere-second, 25
Angstrom, 24
Antineutron, 28–35
Antiparticle, 32
Artificial radioactivity, 102–106
Atmosphere, 41
Atmosphere, radioactive, 130
Atom, 41, 42
Atomic constants, 25–27
Atomic de-excitation, 57–58, 69
Atomic energy units, 25
Atomic excitation, 57, 111
Atomic mass unit, 24–25
Atomic nucleus, 47, 48, 65
Atomic number, 36, 38, 49
Atomic shell, 51–52, 62
Atomic size, 48
Atomic structure, 47–64, 52
Atomic subshells, 62
Atomic weight, 26
Atomic work units, 15–16, 25
Average life, 81
Avogadro's number, 26

B

Barn, 121
Baryon, 33
Baryonic number, 33
Beryllium, 61
Beryllium-7, 90, 108
Beryllium-8, 91
Beta decay, 11, 86–97
Beta decay, light elements, 91
Beta emission, 86–97
Beta (relativistic velocity ratio), 26
Binding energy, atomic, 57–60
Binding energy, nuclear, 68–74
Binding, ionic, 55
Biological half life, 139
Body burden, maximum permissible of Ra-226, 86
Bond, covalent, 55–56
Bone, radium dosage to, 141
Boron, 61
Boron, brain tumor therapy, 122
Boron-10, 83
Boron-13, 86
Brain cancer, 122
Bremsstrahlung, 127
Bubble chamber, 112

C

Cadmium-113, 121
Calcium-41, 20
Calcium-48, 98
Californium-254, 103
Calorie, 132
Calorimetry, 132
Cancer therapy, 116
Capture, neutron, 102, 105–107, 119, 121–122
Carbon, 38, 63–64
Carbon atom, 61
Carbon dating, 101
Carbon-11, 108
Carbon-12, 25, 89
Carbon-12, nuclear states of, 70–73
Carbon-12 thermal neutron capture cross section, 121

Carbon-13, 88
Carbon-14, 100, 106
Cesium-137, 102, 104
Chain reaction, 105
Chamber, bubble, 112
Chamber, cloud, 111
Chamber, free air, 134
Chamber, thimble, 134
Characteristic radiation, 127
Charge, 25
Charge, conservation of, 32
Charge, electronic, 25, 28
Charged particle reactions, 102, 107–109
Chemical behavior, 49
Chemical dosimetry, 132
Chemical properties, 52–56
Cloud chamber, 111
Cobalt-57, 108
Cobalt-60, 96, 102, 106
Colloid, 46
Compton scattering, 129
Conservation laws, 21, 32–34
Contrast media, 100, 129
Control rods, 121
Controlled fission, 105
Conversion electron, 74, 97, 106
Coulomb, 25
Covalent bond, 55–56
Crystal defects, 43–44
Crystalline structure, 12, 43–44
Crystals, 43
Cubic lattice, 43
Curie, 76–77
Curium-248, 103
Current, electric, 23

D

(d, alpha) reactions, 109
(d,n) reactions, 108
(d,2n) reactions, 108
Decay, beta, 11, 86–97
Decay constant, 76, 81
Decay, electron-positron, 31
Decay law, differential, 75
Decay law, integral, 78
Decay, neutron, 11, 30–31
Decay, nuclear, 80, 83–97
Decay, proton, 29–30

Decay, total, 80–81
De-excitation, 57–58
De-excitation, atomic, 57–58, 69
De-excitation, nuclear, 69
Defects in crystals, 43–44
Density, 23
Depth-dose tables, 136
Deuteron, 110, 113
Deuteron-induced reactions, 108
Diagnostic x rays, 100, 128
Dielectric constant, 8
Differential decay law, 75–76
Disintegration, photonuclear, 130
Dose, 131–133
Dose at a point, 131
Dose, calculation of, 136–141
Dosimeter, ferrous sulfate, 132–133
Dosimeter, Fricke, 132–133
Dosimetry, chemical, 132
Dyne, 4

E

e, 78
Effective half life, 141
Electric charge, 25
Electromagnetic force, 5, 7–9
Electromagnetic radiation, 124–127
Electromagnetic spectrum, 124, 127
Electron, 28–35, 48–52, 116–119
Electron capture, 90, 108
Electron, conversion, 74, 97, 106
Electron, free, 45, 63
Electron, interactions with matter, 116–119
Electron, mass of, 24, 116
Electron-positron decay, 31
Electron range, 117–119
Electron range-energy curve, 117, 118
Electron straggling, 117–118
Electron track, 117
Electron, unpaired, 59
Electron, valence, 53
Electron volt, 15–17, 25
Electronic charge, 25, 28
Electronic orbits, 50–52
Electrostatic constant, 8
Electrostatic force, 5, 7–9, 10, 15, 66
Element, 36–37
Elements, second-row, 60–61

Index

Emulsion, 46
Energy, 18–22, 52
Energy, conservation of, 21–22, 33–34
Energy, kinetic, 20–21
Energy level diagrams, atomic, 58–60
Energy level diagrams, nuclear, 70–74
Energy levels, atomic, 56–60
Energy levels, nuclear, 67–73
Energy of a photon, 26, 124
Energy per ion pair, 111
Energy, positive and negative, 19–20
Energy, potential, 21
Energy units, atomic, 25
Energy, units of, 19
Erg, 15
Evaporation, 42
Even-even nuclei, 93
Even-odd nuclei, 93
Excitation, atomic, 57, 111
Excitation, nuclear, 69–74, 96, 97
Excited states, 35, 49
Exponential function, 78
Exposure, 133–136
Extrapolated range, 118

F

f factor, 136
f factor, graph of, 137
Ferrous sulfate dosimeter, 132–133
Fission, 10, 37, 67, 102–105
Fission, controlled, 105
Fission, induced, 103
Fission, spontaneous, 103
Fluorine, 48, 53, 54, 61
Fluorine-18, 109
Foot-pound, 15
Force, 3–11, 13–14, 23, 25
Force-at-a-distance, 3
Force, contact, 3
Force, electromagnetic, 5, 7–9
Force, electrostatic, 5, 7–9, 10, 15, 66
Force, gravitational, 5–7, 14
Force, intermolecular, 42, 65
Force, kinds of, 3, 5
Force, magnetic, 5, 7–9
Force, nuclear, 5, 10–11, 65–66
Force, units of, 4
Force, weak, 5, 7, 11
Free air ionization chamber, 134

Free electron, 45, 65
Fricke dosimeter, 132–133
Fundamental particles, 28–35

G

Gamma rays, 86
Gamma rays, definition of, 127
Gases, molecular, 41–42
Gases, noble, 41, 53
Gold-198, 103, 106
Gravitational constant, 6
Gravitational force, 5–7, 14
Grenz radiation, 126
Ground state, 49

H

Half life, 79–81
Half life, biological, 139
Half life, effective, 141
Halogens, 53–54
Helium atom, 60–61
Helium-3, 91
Helium-4, thermal neutron capture cross section, 121
Helium-5, 91
Hydrogen, 25, 37, 48, 119
Hydrogen atom, 56
Hydrogen binding energy, 56–60
Hydrogen-2, 83
Hydrogen-3, 138

I

Induced fission, 103
Infrared, 124
Ingested radioactive material, 136
Insulators, 62
Integral decay law, 78
Interactions, proton, 110, 116
Interatomic spacing, 126
Intermolecular force, 42
Iodine-131, 77, 103
Ion, 42, 29, 66
Ion pair, energy per, 111
Ionic binding, 55
Ionic solution, 54
Ionization, 111
Ionization, minimum, 113, 116, 117
Ionization, specific, 113, 115
Ionizing radiation, 125

Iron-59, 106
Isobar, 40
Isobaric family, 40, 93–97
Isomer, 68, 97
Isometric state, 68
Isotone, 40
Isotonic family, 40
Isotope, 36, 39–40

J

Joliot-Curie, 102
Joule, 15

K

Kinetic energy, 20–21

L

Lead, 206
Lead, photoelectric absorption, 128
Lead-206, 84–85
Lead-208, 84
Lead-210, 85
Length, units of, 24
Lepton, 32
Leptonic number, 32
LET, 113, 115, 116, 120
Light, velocity of, 26
Light, visible, 124
Liquid drop model, 66
Liquids, structure of, 42
Lithium, 61
Lithium-6, 83
Lithium-7, 68, 90

M

Magnetic force, 5, 7–9
Mass, 4
Mass, electron, 24, 116
Mass, fundamental particles, 29
Mass, neutron, 25
Mass number, 38
Mass, proton, 25
Mass unit, atomic, 24–25
Mass, units of, 24
Maximum range, 118
Medical uses of radium, 85–86
Mercury, chemical properties, 49
Mesons, 65

Metalloids, 64
Metals, 45, 62
Metals, structure of, 45
Metastable states, 68, 70, 73, 97
Metric prefixes, 24, 25, 77
Micron, 24
Microscopic physics, 23–27
Microwaves, 125
Minimum ionization, 116, 133, 117
Molecule, 12, 41, 48, 52–56
Molybdenum-99, 97, 103, 106
Multicrystalline substances, 44, 45

N

(n,p) reactions, 106–107
N/Z ratio, 31, 83, 92, 104–105, 108
Natural radioactivity, 98–101
Naturally produced radionuclides, 100
Neutrino, 28–35
Neutron, 28–35, 119
Neutron capture, 102, 105–107, 119, 121–122
Neutron decay, 11, 30–31
Neutron mass, 25
Neutron number, 38
Neutron-proton collisions, 120
Neutron scattering, 119–121
Neutron therapy, 122
Neutron thermal, 120
Neutron thermal capture cross section, 121
Newton, 4
Nitrogen, 61
Nitrogen-12, 88–89
Nitrogen-14, 83, 106
Noble gases, 41, 53
Non-ionizing radiation, 125
Nuclear bomb, 105
Nuclear de-excitation, 69
Nuclear energy levels, 67
Nuclear excitation, 69–74, 96, 97
Nuclear fission, 10, 37, 67, 102–105
Nuclear force, 5, 10–11, 65–66
Nuclear level diagrams, 70–74
Nuclear models, 65–67
Nuclear orbital diagrams, 68
Nuclear reactor, 105, 107
Nuclear reactor control, 121
Nuclear shell model, 67

Index

Nuclear structure, 47, 65–74
Nucleon, 29
Nucleon orbits, 68
Nucleus, 47, 48, 65
Nuclide, 38–39

O

Odd-even nuclides, 94
Odd-odd nuclides, 94
Ohm, 9
Orbital model, 52
Orbits, electronic, 50–52
Orbits, nuclear, 67–70
Organic compounds, 45
Original nuclides, 98
Oxygen, 25, 55–56, 61

P

Pair production, 130
Particles, fundamental, 28–35
Periodic table, 37, 49
Phosphorous-30, 102
Phosphorous-32, 106–107
Photoelectric absorption, 128–129
Photon energy, 26, 124
Photon, interactions with matter, 122–130
Photonuclear disintegration, 130
Physical quantity, 23
Pile, nuclear 125
Planck's constant, 26–27
Planetary model, 49–52
Plutonium-239, 104, 105, 107
Polonium-210, 85
Polymerization, 45
Positron, 28–35
Positron emission, 90, 108
Potassium-40, 98
Potassium-41, 20
Potassium-42, 106
Potential energy, 21
Pound, 4
Practical range, 118
Promethium, 93
Proton, 28–35
Proton decay, 29–30
Proton interactions, 110–116
Proton mass, 25
Proton track, 112

R

Rad, 131
Radiation, characteristic, 127
Radiation, electromagnetic, 124–127
Radiation, Grenz, 126
Radiation, ionizing, 125
Radiation, non-ionizing, 125
Radio waves, 125
Radioactive atmosphere, 130
Radioactive material, ingestion of, 136
Radioactive series, 100
Radioactive waste, storage of, 104
Radioactivity, artificial, 102–106
Radioactivity, natural, 98–101
Radionuclides, naturally produced, 100
Radionuclides, original, 98
Radium-226, 84–86
Radium-226, dosage calculation, 142
Radium-226, hazards of, 86
Radius of fundamental particles, 29
Radon, 85
Range, 113–116
Range, electron, 117–119
Range-energy curve, electron, 117, 118
Range, extrapolated, 118
Range, maximum, 118
Range measurement, 114
Range, practical, 118
Reaction, chain, 105
Reactions, charged particle, 102, 107, 109
Reactions, (d,alpha), 109
Reactions, (d,n), 108
Reactions, (d,2n), 108
Reactions, deuteron induced, 108
Reactions, (n,p), 107
Reactors, control of, 121
Reactors, nuclear, 105, 107
Relativistic kinetic energy, 21
Relativistic mass increase, 116
Relativity, 26
Rest mass energy, 33
Roentgen, 131, 133

S

Scattering, Compton, 129
Scattering, neutron, 119–121
Semilogarithmic plot, 78
Shell, atomic, 51–55, 62

Slug, 20
Sodium chloride, 54
Sodium-24, 106
Solids, amorphous, 43
Specific ionization, 113, 115
Spectrum, electromagnetic, 124, 127
Spontaneous fission, 103
Stability, light elements, 91
State, ground, 49
States, atomic, 49–52
States, excited, 35, 49
States, metastable, 97
Stopping power, 113, 115
Storage of radioactive waste, 104
Straggling, 113–115, 117–118
Strontium-90, 102, 104
Structure of crystals, 12, 43–44
Structure of liquids, 42
Subshells, 62

T

Technecium-99, 72–73, 93, 97
Therapy, cancer, 122
Therapy, neutron, 122
Thermal neutron, 120
Thermal neutron cross section, 121
Thimble chamber, 134
Thorium, 100
Thorium-232, 98–99, 103
Thyroid gland, beta dosage to, 139
Time, units of, 24
Total decay, 80–81
Track, alpha, 113
Track, deuteron, 113
Track, electron, 117
Track, proton, 112
Transformations of fundamental particles, 29–35

Triplet production, 130
Tritium, 137

U

Ultraviolet, 124
Unpaired electron, 59
Uranium, 37, 99
Uranium series, 84–86
Uranium-235, 98–100, 104, 105
Uranium-238, 98–100, 103, 106

V

Valence electron binding energy, 125
Valence electrons, 53
Velocity, 23
Velocity of light, 26
Visible light, 124
Volt, 8, 17

W

Water, 48, 55
Wavelength, 123
Weak force, 5, 7, 11
Weight, 3, 4, 5
Weight, atomic, 26
Work, 12–18
Work units, 15–18
Work units, atomic, 15–16, 25

X

X radiation, 126
X radiation, diagnostic, 100, 128
X-ray contrast media, 100
X rays, definition of, 127
XU, 24